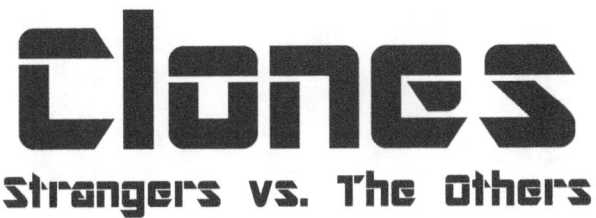

Strangers vs. The Others

Giovanni Rocca

MR Comics & Art • Andover, Massachusetts

MR Comics & Art
5 Binney Street
Andover, MA 01810

Published 2013

ISBN 978-1-4675-9434-9

Book design by Kevin A. Wilson
Upper Case Textual Services
Lawrence, MA

Cover art by Giovanni Rocca

Introduction

Science embraces the theory that strangers have visited planet Earth in the past, and someday we will encounter these species from outer space: UFOs, ETs, aliens, superior and advanced intelligences beyond our imagination. Meanwhile, governments and organizations spend an incredible amount of money and time to come up with some answers, but the outcome is nothing of value. They cannot show the existence of aliens beyond fiction and imagination. The search for aliens continues for generations. Some go to school from elementary to college or university to study and learn bogus material. They know that there is no proof for what they teach, but who cares as long as money comes in. If aliens or ETs existed, we would not be here; they would exterminate the human race the same way humans do with rats, bugs, insects, and all kinds of living things that humans are above. The human race is stupid and unkind, selfish and power hungry, delusional and ignorant. Above all, it is malicious and destructive. In reality, there are no words to explain humans. Scientists think that by investing their time in unknown things their knowledge will expand, but the more they learn the more they need to learn.

Science and medicine go hand-in-hand. Researchers

practice and practice until they drop. They try to find cures for diseases from the beginning of time. For the simple common cold, no cure has been found after five or six hundred years of manipulating human cells. They manufacture drugs from natural herbs or make synthetic drugs, but the virus and germs are still around the human body and growing. Not only can they not cure colds; they do not know how to begin to understand the problem. But they love to play god. Man is already in trouble with himself. Every year that goes by, thing get worse, from medicine and science to new technology. The elderly have a hard time using and understanding it, while young people go crazy using new devices with new technology. You can see the competition on making computers, television, cellular phones, and games, and technology from tubes on radios and televisions to plasma, LED, LCD, 3D, FiOS, and digital signals. Who knows what they will come up with next? Again, man is ahead of himself with dreams and excitement for new ideas for a comfortable life. It will get better or it will get worse for the human race.

I believe that it will get worse every day for the human race. It will not get better as long as man tests new chemicals and unnatural ways with animals and plants. And on top of all that, they experiment on the human body. Morality is the key. If man loses that, he will expire.

In the Old Testament book of Genesis, God warned man, "Do not eat of the tree of knowledge." God tells man that if he disturbs creation, man will suffer the consequences in his body and soul, but man disobeys God for his selfishness and curiosity, experimenting and testing the unknown. It will lead to self-destruction.

This is the third installment in the Clones trilogy. It will be a conflict between good and evil; the Strangers will battle the Others. The humans are the objective here; they are the problem of the planet Earth and of the universe. The question is, are humans made by aliens or aliens by humans. We will discover what went wrong in the creation of the clones and why man cannot stop putting his nose where it does not belong. The end is at hand. We know that it is real. The only question is when.

In this book, I am giving my own feelings as a work of science fiction. You can think and judge for yourself what is real and what is fiction. What would be the possibility that the aliens we believe to be from outer space are really in our own back yard, in ourselves. As good and evil are within us, so strange things of the imagination are too. You can make anything you want as long you have an imagination.

I say again and again, I do not believe. I know there is no extra-terrestrial intelligence out there beyond the planet Earth. If it were, the human race would be used and consumed by them, just like we do with any of creation we see and destroy what we don't like.

Conflict between the clones will take place in the future, far out into the universe, for the hope of humanity and the planet that is most important, planet Earth. The Strangers—the clones that Gianni met—gave Gianni a massage to be delivered to the leaders of the world, stating that the planet Earth was in danger of destruction and to take their warning seriously, because time was short. The leaders tried to obey the message the best they could, but they were confused. Even though the message was clear, they did very little. The Stranger appeared during the conference with Gianni to warn them of the outcome: the end of the world and the human race.

The Others traveled many light years to arrive at the Earth's star system. From their mother ship, they observed below on the planet Earth. They saw a family working with animals and planting seeds. They saw two men walking with a herd of lambs, and then one of the men took a jawbone and hit the other man on the head, killing him. The Others saw the act and were pleased. The Others used that man for experiments for the future of the human race. The Others are the clones gone bad, pure evil. Unlike humanity, which went from perfection to imperfection, in whom is both good and evil, the Strangers are all good, and they need to save the Earth, their new planet for their survival. They are the guardians of the universe.

FIRST CONTACT

Space—a vast, dark emptiness with stars and galaxies, gases and dust, infinite in time and distance. No mathematical measure exists for man and his little knowledge of the vast creation called the universe.

A battle is taking place in space between the Strangers and the Others, good and evil. The Strangers are fighting for their planet and their survival.

In 1996, the Strangers came to Earth to investigate their existence, to learn how they came about, and to know the truth of their makeup. They wanted to learn the history of their past to maintain the security of their future. After thousands of years of searching and tracking their DNA, they arrived at a location in Italy, where they sensed the energy and the source they were seeking.

"I sense that he is the one," clone Primo tells his brothers.

"We sense that too," his brothers reply.

"Let us descend to him and make him comfortable," Primo suggests.

They communicate telepathically. They descend to a man sitting under a fig tree, resting from a hard day of

1

work. As the Strangers float about one meter above the ground, the man opens his eyes and sees a shining ball. He hears a mellow and vibrant sound. For a second he goes blind, and then his eyesight comes back. He witnesses something strange and fantastic at the same time.

"Oh, no! It is true! Aliens are real," the man thinks.

The Stranger floats towards the man, and then two more of them came out of the ball. The ball changes colors: light blue, green, opaque yellow, white. The gigantic figure approaches the man and says, "Do not be afraid. I am a friend. I will not harm you, Gianni!"

"He called me by my first name," the man thinks. He becomes tranquil.

The Strangers and Gianni departed the Earth in the blink of an eye.

The zero year for mankind. It is a glorious day with the sun shining upon the Earth. A family is working, minding their sheep and farming. Two men are walking and shepherding the sheep. One of the men takes a jawbone of a donkey and strikes the other man on the head, killing him.

Above in a spaceship, someone is watching the act below. One of them exclaims, "That is very good. He is the one we need!"

The man down below hears a thundering sound and begins to run. He runs as fast as he can from that place of death, and then he falls on his face.

Earth, the blue planet, full of life, the home of humanity and all living things. But something is very wrong.

Instead of apes, the alien finds humans with brains. The Earth has been visited a multitude of times, especially by the aliens from the deepest and most remote places in the universe. Some left good signs; others left evil signs. All were intended to help man advance. But right from the beginning of mankind, everything he put his hands on turned to waste. The aliens know that mankind is weak, and he is a follower. Humans are cowards, liars, and deceivers.

The aliens came to Earth to search for resources they can use for their own benefit. They came to extract all minerals from the planet, without care or respect for anything. They enjoy destruction and chaos, and their goal is to avenge what man has done to them.

Inside the mother ship, the Others were cloned. The first was named Me. He is the head of all the Others. Next to him are the twenty clones made after him. They are looking at a large screen. They see planets and galaxies. They are looking for planet Earth. They know that it came to an end. They need to know the exact location of Earth and its galaxy, so they can find the source of energy that they were detecting before the Earth was destroyed. Every time they departed Earth, they saw the blue ball passing by them towards the Earth.

"It is strange that the planet Earth does not exist anymore," Me exclaims.

"Maybe we have gone to another dimension," one of the brothers responds.

"Something is wrong here."

They check the space map on the screen. By referencing the location of the stars, they realize that they are in the Earth's solar system, but they do not see the Earth. They arrive near the planet Dopa, fifteen planets beyond Pluto, about 12 billion kilometers on the other side of the solar system. They look the screen, at the vast space that separates the planets from one another.

"There ... a new planet?" Me says to his brothers.

"What do you see, Me?"

"Look at that planet! Is that a different planet, or a new planet?"

Me zooms in on the planet, but he cannot see much. The new planet is covered with a thick atmosphere. The new planet is located in the same orbit that the Earth previously occupied.

"We need to get closer and send a probe, so we know what's there."

They send a probe to investigate, but the probe is repelled with great force from the planet's atmosphere. The Others are puzzled and surprised. They look at one another. Then they turn back to the large cyberspace screen. They see nothing.

"What was that?" Me asks.

"The planet seems to have a protective energy field around it," one of the others replies.

"It is possible," another said.

"Do you feel that maybe the planet is inhabited?" Again, Me asks questions beyond his knowledge. For the first time, he is confused.

"The best thing is for us to return later in the space-time continuum." Me wants to change course. He feels that if they come back later, they will have a better understanding. The Others depart at light speed, leaving behind the green planet.

The green planet changes to a clear sky, looking like a green marble ball. Slowly, a sphere moves up towards the dark space and circles the planet three times. Then more spheres fly up into space, moving outwards in all directions into the universe, traveling to other planets to investigate what just occurred. The large sphere moves away from the green planet faster than the speed of light. It disappears in the blink of an eye.

As for the Strangers, Primo and his two brothers are vigilant. They know that it was a force beyond anything they have ever known, and it will come back to their planet. They will check every galaxy and every planet for the destructive, dark, evil energy.

"We will warp time and travel into the future. There we will possibly find knowledge of the destructive force." Primo is talking with his brothers. They arrived at a galaxy millions of light years from their home planet. The look and see that this galaxy has three arms that spin clockwise and three counter-clockwise. The galaxy has millions of suns and trillions of planets and moons.

"This is a grand galaxy, and it has a strange shifting movement, different than our galaxy," Primo is thinking. At the same time, the other brothers are thinking the same thing, looking at one another.

"We will name it Geminius for the two suns. It will be our starting point for the search," Primo expresses to his brothers. Primo and his brothers continue the search for the dark energy.

THE FINDING

On planet Earth, governments from America, China, and Europe experiment with new avenues of research, cloning plants and animals. Slowly, they progress and get better with time. They continue to experiment, creating larger and more complicated studies. After many years of experiment with plants and animals and all kind of living things, they make rapid advancements in science and medicine. Right or wrong, they keep going non-stop with no compromising and no penalties to worry about. Their mistakes and misshapen creations are protected by the governments.

America, Europe, and China begin to experiment with cloning plants and animals. They progressed incrementally. Essentially, their experiments in cloning plants and animals tripled the food supply for civilized countries. Experiments began when the first man planted seeds and split them to grow different fruits and vegetables. Experiment after experiment, humanity evolved in knowledge of how to manipulate nature in its own way, with few consequences or penalties for their mishaps. The 20th century saw experiments of all kinds increase to the point of no return. Many mistakes and deaths occurred.

Around 1930, the German government put together a group of scientist to study cloning to produce the perfect race of humans. To keep the outcome secret, during that time many people died off the record. The Chinese government was also studying cloning behind closed doors. No one in the world knew of their secret work. In 1981, the first space shuttle orbital test flights occurred, with operational flights beginning in 1982 at continuing until 2011 for NASA. But other countries continued. In 1976, the Soviet Union had space stations—*Salyut 5* and *Almaz*—used for military and civilian purposes. Only a few scientists knew that the United States launched their first advanced spaceship and that the European Union also sent their spaceship into orbit for the purpose of H.C.T (Human Cloning Experiment).

The American government with NASA sent secret vessels into space without their citizens' knowing what was in the space shuttle and what was really going on up in space. The government told the citizens and the world that they were just performing experiments and tests. Again, lies and more lies to keep people quiet and to encourage contributions to NASA's study of space and to America being first in the space race. The American people did not care about the truth; for the moment, it was fine with them as long as the US was number one.

NASA was successful in sending space shuttle after space shuttle, until 1986 when seven crewmembers died in the space shuttle *Challenger* disaster. It was the biggest and the most shocking visual spectacle for the nation.

The US government and NASA informed the world in a report that contained a false transcript of the incident that occurred on the space shuttle. They claimed it was ice the formed on one of the solid rocket boosters that caused the explosion. The case was closed, as people accepted the explanation of ice on the booster. Congress investigated the matter, but no one questioned or challenged the government or NASA.

Justin, a scientist who worked on the space shuttle, knew the truth about placing regular people on the shuttle with some degree of professionalism. They were people with very little experience in space walk and repairs. They were more guests than astronauts. Justin told the head scientist in charge of the mission not to launch that particular shuttle into space.

"I believe if this mission takes place many things will go wrong. We need to think this situation through very carefully! I am a scientist, I know, but my honor and morality will not allow me to participate any longer. It is unjust to the American people and the innocence of these blind astronauts!"

Justin gave the same information to his partner. He was worried and afraid of the outcome. His friend looked at him, also worried. He shakes his head.

"Justin, we cannot stop working on this mission or say anything about it. We are scientists, not informers! Even if we do inform the government, then what? They will do whatever they want, you know! We have no power to change their minds about the mission. Besides, we have family. Who knows their fate would be if you spilled the beans!"

"You are right. I should be quiet."

And so the scientists continue the work in preparing the shuttle for takeoff. Justin is very nervous and stressed. He cannot concentrate on the job. His friend sees how Justin is acting and how nervous he is. He is afraid that Justin will talk about the problem he encountered.

After a week of intense work, the shuttle is ready to be launched. The previous week saw a grand celebration for the astronauts and their mission. The country is excited with the great news that hit the whole world. America is again number one.

Justin is with his scientist friends at the dining room at NASA, eating and conversing about the shuttle and the astronauts. Out of the corner of his eye, Justin sees a figure. It appears to be is wearing a black suit and dark glasses. The figure appears to be speaking into a transmitter.

Justin panics. Slowly, he gets up and starts to walk to his room.

"Friends, I'm tired. I will retire for the night. See you in the morning."

THE DEATH OF JUSTIN

Justin arrives at his room and enters, closing the door behind him. He then opens the closet and takes out paper and a fancy pen. He sits at his desk and begins to write. After an hour, he finishes writing. He places the note in a briefcase with other documents and locks it.

The day arrived for the shuttle launch. All the scientists watch the big screen, which is showing the famous shuttle and the lineup of the astronauts, smiling and excited for the mission. They enter the shuttle, and after the countdown, the shuttle lifts off. It rises into the sky. After slightly more than a minute of flight, there is an explosion. Shock and chaos ensue as the people look up in the sky, witnessing the unbelievable: the disaster of the space age and the future of space missions.

At the same time, Justin's friends are worried about him. They go to see him. They knock on the door.

"Justin? Justin, it's me, Steve! Open the door, please!

There is no answer from inside. For a moment, it is quiet, with Steve and the others in silence. Steve again knocks on the door.

"Open the door, Justin!"

Again, no response from Justin. They force open the door and see a gruesome picture. Justin body is hanging from a ceiling bar.

They call for help, and two government agents arrive at the scene. They tell them not to touch anything around the room and for them to go.

"Thank you for getting us. We will leave this to the investigators. You can go now. We will take from here."

After the investigation, two government agents visit Justin's family. The agents knock on the door. The door opens and a lady appears. The agent looks at her.

"Mrs. Forest?" asks the agent.

"Yes, that's me," she replies.

"We are government agents."

Mrs. Forest looks at them in anticipation of why they are there.

"We have bad news for you."

She looks, waiting for them to tell her the bad news.

"May we come in?" they ask.

"Yes, you may," she replies, very nervous.

"Your husband is dead. He committed suicide."

She is stunned by the bad news. She sits on the couch looking at them with sad and angry eyes.

"My husband would never do such a thing! He loved life! And his family!"

"We are sorry. Here are his office items."

The agents leave the Forests' home. Mrs. Forest closes the door.

The next day Mrs. Forest and the children are in the living room, talking about their father.

"Your father was a good man! He would never commit suicide. He loved God! You know that he loved you! He was a man of high standards and a moral way of life."

"We know, mom. Father always said to live right with God ... not to be a liar or a criminal." As he finishes, they embrace.

Three days pass. They have Justin's funeral. Family, friends, and government officials are present, as is Justin's friend Steve. After the ceremony, Steve approaches Mrs. Forest and kisses her on the cheek.

"I am so sorry at Justin's passing. If is not inconvenient, I would like to visit with you all," Steve whispers.

"No trouble at all, Steve," she answers.

"In a day or two, if that would be okay."

The government and NASA continued the investigation and reached a conclusion on the shuttle incident. They need some conclusive fact or anything to tell the world what occurred. The president and scientists from NASA meet to discuss it. The scientists come up with false information to deliver to the people of America and the world.

"I am going to give a news broadcast to inform our people and the world about the facts and the truth of the shuttle incident. I will also take some questions," the president says.

"If you must, sir," the head scientist replies.

The president and his advisers are in the Oval Office. There are many journalists ready for questions. The president comes in, takes the microphone, and begins to deliver the message.

"Ladies and gentlemen, today I will give you the report

from the committee set up to investigate the shuttle disaster."

The president begins to read the report. The reporters are standing and listening to the president.

"The report concluded—after many hours of investigation—that a failed O-ring on one of the solid rocket boosters caused the explosion." The president looks sad and angry at the same time. He pauses for a moment.

"I am sorry and heart-broken for those who lost their lives. It will not happen again."

The president finishes delivering the news and walks away without taking questions. The president never knew the truth.

Two days pass before Steve visits the Forest family. Inside the Forests' home, he and Justin's family are conversing.

"How are you coping with the death of Justin?" Steve asks.

"Fine, Steve, but it is hard. With time we will adjust," she replies. "How about you? Are you okay? After all that's happened with the shuttle disaster and Justin's death?"

"It is a lot to take in right now," he says with a shaking voice.

"Would you like some coffee?"

"Yes, please, with some milk and sugar." Steve is very upset and stressed. He drinks his coffee with tears in his eyes.

"May I have a glass of water?" Steve requests.

"Certainly," she says, with sorrow and pain for the lost of her husband.

As she walks towards the kitchen, her son follows her.

"Mom, Steve is really sad and nervous."

"Here, Steve. Drink this water instead of the coffee," her son says.

Shaking, Steve drinks the water slowly with his head down.

"What is it, Steve? You are so nervous."

"I have something to tell you," Steve says quietly.

Mrs. Forest and her children look at Steve anxiously, waiting for Steve to tell them the news.

"What about, Steve?" she asks.

"It is in regards to Justin and his last words to me before he died."

"Is it about him and NASA?" she asks nervously.

"Yes. Justin said to me that he discovered defects on one of the tanks carrying the space shuttle and the astronauts into space." He stops for a moment, breathing heavily.

"He knew more than I did. I told him that we should mind our own business. I said that we are not informants ... we need to be careful because we have families. We don't know what the government would do to us if they knew the truth." Steve is interrupted by Mrs. Forest.

"You are telling us that Justin found out something that was very wrong, and you stopped him from confronting the head scientist?" she asks angrily.

"Yes ... to protect you and my family," he replies with remorse.

"The government knew what Justin discovered and kept hidden? This is outrageous, Steve. It is bad for the country and the human race." She begins to cry.

"I am so sorry, but I know Justin's findings will be exposed ... I promise you!" Steve embraces her and cries with her.

"Do you think the government is responsible for Justin's death? Are NASA and the US government involved in the destruction of the shuttle?" she asks Steve.

Steve hesitates before responding. "I don't know if they are responsible for the death of Justin. But what Justin said to me suggests that, yes, our government and NASA are responsible for that accident!"

"Why? Why? This is unbelievable! I find it hard to accept!" she exclaims with pain and anger.

"Whatever Justin found is gone now ... forever. There is no paper trace. Only government secrets," he says with his head down.

"This is dangerous, Steve. What are we going to do?" She worries about her children and herself.

"I beg you all to be vigilant. You never know what they will do to protect and hide this evil act," he says.

With that, Steve embraces them and goes away.

NASA

At NASA headquarters, the head scientist and six others, including engineers and mathematicians, meet to discuss the shuttle and the problem that occurred. They are watching a large screen that show a video of the terrible explosion. The head of science is talking. He points with a laser to the location of the explosion.

"Please, watch this film with attention," he instructs.

He plays the film many times back and forth. Then he stops right before the explosion, freezing the picture. They engage in conversation, discussing the mechanical problem. An engineer stands and takes the laser light and points at the fuel tank.

"Observe here, gentlemen. Please, let the film advance slowly."

They look at the film frame-by-frame. He stops the film at the moment of explosion.

"Do you see what I see? The explosion occurs inside the orbiter, not in the tank!"

They look at the film closely.

"You are right! But that is impossible. The report indicates that it was a fuel tank malfunction," one of the scientists exclaims.

The mathematician looks at all the others with suspicion. "I don't know anything of this matter. If you know something, better say so right here!"

They look at each other, but no one speaks.

The next day at Justin's house, Mrs. Forest opens the package given to her by the agents. She is alone, sitting on a chair. She begins to open the package of Justin's materials from his office. One-by-one, she removes articles. Then she opens a notebook. The contents are Justin's notes from work ... nothing of value to her. She places the notebook on the coffee table and goes into the kitchen to get a glass of water. Then she continues to look at the book. She comes to blank pages. They have no writing on them. They are simply white pages.

She drinks a little water. She is crying her hands are shaking. She spills water on the blank pages.

"My God what is this ... What is happening here?" With her eyes wide, she looks at the pages that got wet. They begin to transform. Writing appears on the white pages.

"It looks like magic! From blank pages, now I see writing ... Oh, my God, it's Justin's handwriting! A message from him!"

It begins to make sense, little by little. The empty pages are full of Justin's writing. It becomes a message from Justin to his wife. She drinks some water and begins to read the message. She is nervous, sad, and anxious.

My dear and loving wife,

I hope that this note finds you and the children well. If you are reading this message, that means I am dead. This note is for you and for you alone. Be vigilant. Do not converse or say anything in regard to this. If this note gets out, you and the children will be terminated by our government.

I am watched by government agents. They know that I have information on the impending shuttle disaster. In fact, I know that our government and NASA are working together to cause the shuttle disaster. I tell you the explosion will be purposely achieved by them, and then followed up by a falsified report to the world. This will be done so they can have more money for scientific space missions and have people from all over the world take an interest in the adventure. The US wants to be first; they will do whatever it will take to accomplish their goal to be number one, even if innocent people die for the outcome. I heard some government officials and some NASA technicians and scientists making plan for the shuttle extermination. These are their words:

"We must make it happen, to make things right with the people of the world. If we have a disaster with the shuttle, we will continue our research. If we do not, NASA will dissolve and we will be out of work for good. We have nothing to report from our missions, our exploration of space. People and companies will be tired of no news from NASA ... nothing ... nothing ... again and again. We cannot go on. We have to do something quickly.

We can place a small amount of plastic explosives the size of a golf ball on the orbiter. It will ignite under heat so there will be no trace of any kind. With this we will guarantee the continuation of our research."

They went forth with this evil idea. I attempted to communicate this information to Steve, but I worried for him and his family. I had only a little time to speak with him. I did not tell him much. The FBI agents had me under surveillance and discovered that I spoke with Steve. They kept a watch on me. They came to my room and warned me that if they found any information about the shuttle I would be in deep trouble.

I know that I have a little time, so I got my notebook to write this note for you, in case they terminate me. I am using a special ink that I invented. It is water-soluble. After you read the message, just pour water on it and it will disappear. My love, with these words I say goodbye. May God watch over you and the children. I love all of you! If I do not return, you will know for fact that our government terminated me.

Your loving husband,
Justin

After she reads the note, she burns it. Now the note is gone with Justin, but the information is with his wife. She knows the truth and she will keep it for as long as she lives.

The US government, NASA, and other nations continue their missions to space. They send many more shuttles with different payloads of material. Many of the non-shuttle missions were un-manned. Ten to twenty space flights a year were launched from Europe, Russia, and China. Russia had a space station, which they were loading with materials to build a larger station for study and research. The US also built a space station for research in technology and advanced medicine with animal, plants, and human cells. They report that the research in space is for the benefit of the humanity and the planet Earth.

Donations pour in from all corners of the world and from US citizens. Then, in 2003, after many delays, NASA finally sends a shuttle with seven astronauts on board with an important payload. They say it is the most important of all missions. On board the shuttle, the astronauts are very excited and optimistic for the mission. But once they are in space deploying the payload, they discover that in the package is a nuclear arsenal and human body parts for experiments. The astronauts meet in the crew cabin to discuss what they have found.

"We can't let this happen! It is immoral and evil to have a nuclear space station! And we need to report the study of humans using body parts."

The astronauts do not know that the space shuttle is bugged. NASA hears everything they say. The astronauts were not aware of the problem with the shuttle.

NASA communicates with the astronauts. "Good morning! How do you feel?"

"Well, thanks. Are we coming home today?" they reply, floating in the cabin.

"We are doing what we can to make it possible for

today. We have a little problem that may affect reentry," NASA answered.

"What is the problem?" the commander asks.

"The shuttle appears to have one or two wing tiles loose."

"Can we fix it from here?"

"No, you don't have the required adhesive," NASA replies.

"Then what?" the commander asks.

"We are investigating the problem and trying to come up with a solution. We will transmit again in a few hours." They close communications.

Mission control is unaware of the transmission that just occurred between NASA and the astronauts because it was done from a secure office. No one in the center or any other part of the country knew the truth.

Two hours pass. The astronauts have opened communication with NASA again. They think they have fixed the problem and are ready to go home.

"Columbia, this is Mission Control Houston. We are back with good news."

"Great! What we need to do from here?" the commander asks.

"Nothing! The reentry will take place in an hour! Good luck!"

As they disconnect they prepare for the reentry to Earth. That was the last of them. The space shuttle *Columbia* disintegrates during reentry. The US government reports that the disaster occurred because of a loose tile on the wing. It stops all mission for a while until they inspect all the problems that the remaining shuttles may have.

Over the next few years, the government and NASA send secret payloads to the space station high up above the Earth. Then they go on TV to report that they have good news with regard to the shuttles and space exploration for the good of humanity and the world. The public again is told another lie. Donations pour in from stupid, brainwashed people and big companies who support the lie.

The time comes when NASA finally built an advanced space shuttle called *Adam*. Europe built a similar spaceship called *Eve*.

Years pass from the time of greatness. The world becomes a hellish place. It descends into chaos after the economy falls. There are no jobs. Money is not valuable or useful. And then revolts, wars, pestilence, and diseases come. Humans fight for water and food to stay alive. Governments begin to impose order, placing curfews on the people.

VISION

THE FUTURE

The Strangers come and take Gianni to their planet to see the future of mankind. The Others also come to change the path of mankind by using the first evil man, Cain. They will use him for their gain and the purpose of human destruction. And thus, the end came to the human race and the planet Earth. The Strangers are the guardians of the universe. They make sure that the dark energy that they sense will not reach their tranquil home planet, the green planet.

"Primo, we need to maintain order and security for our home and family," C=2 urgently expresses to his brothers telepathically.

"You are right in this regard, brother! It is natural to be cautious after the feeling the dark energy," Primo replies to C=2. He too senses the need to secure the planet.

"It is possible that the dark energy is close to our home. We shall investigate our galaxy!" brother C=3 adds.

"Yes, brother. We will battle them to protect all that we have!"

Communicating telepathically without losing any time, Primo departs at light speed into the universe, protected by a sphere for a long voyage investigating their galaxy and the surroundings. From planet to planet, millions and millions of kilometers, from corner to corner in the blink of an eye, he stops and goes, checking for the dark energy in his home galaxy, the Milky Way. He is vigilant and unwavering in his investigation.

Primo senses nothing after 10 minutes of traveling. In reality, he travels around the galaxy for 1000 years of non-relativistic time. He returns home to the green planet and slowly descends to meet all of the clones waiting for him. He gathers all of them, all 144 million.

A large sphere begins to develop around them. Primo, C=2, and C=3 are near each other. The three of them are floating above the others. Primo bows his head and the sphere changes from sky blue to the dark of the universe. It shows the travels of Primo, all the planets he visited in all corners of their galaxy, the Milky Way. In that vast galaxy, they see no other life and feel no dark energy. They are all comfortable with the sense of security. After the great tour of their galaxy and beautiful green planet, Primo begins to explain the possibility of the origin of the dark energy.

"My loving and peaceful family, as you see I traveled from corner to corner of our galaxy. The dark energy is not here. It is my concern that they will find us. When that occurs, we shall battle for our survival and to preserve our planet of good and peaceful kind for this planet, our galaxy, and the universe."

Then C=2 continues the instructions. "As you know, I feel that evil energy around our planet. The way we will learn more is to communicate with one another when we sense anything strange."

All the clones are stressed and look at one another. They all feel their security and lives in danger.

"The three of us will depart. We will be vigilant for our planet and our galaxy! We will travel in a time warp to the past, present, and future as the guardians of this the universe," C=3 explains.

In an emotional telepathic communication, they express their hopes of finding this great dark evil energy and returning home safely.

The three of them fly away in the blink of an eye into the universe. They travel to three different location of the Milky Way.

"C=2, I will take the A position. You will be on the B position, with C=3 at the C position," Primo communicates to his two brothers in while traveling at the speed of light. "By triangulation, we will have the positions we need to observe our galaxy with 500 billion stars and planets, covering 225,000 light years from our own planet."

Primo, C=2, and C=3 coordinate the plan.

Six light years from the home of Primo, the Milky Way, the Others are engaged in the occupation of planets and moons for the extraction of materials needed for their survival.

"This planet is very different from the planet Earth! We need a living planet for our survival," exclaims Me.

From the inside of their planet-sized ship, they explore

new planets to extract minerals, use the inhabitants, and then destroy them. That is what they do. From planet to planet, they travel at light speed, sending probe after probe. The planets with no special materials that they need are destroyed.

The Others need material continuously for their existence. They need gold, silver, copper, and all minerals that the Earth produced. If they do not acquire them, their existence will cease.

From a long distance, they sense a large planet, similar to Earth. In an instant, the screen shows a large planet with an atmosphere 6.6 trillion parsecs away. The distance does not affect them. They can arrive in minutes.

"This is not the first. We encountered many just the same as this one," Me says to the other clones.

The second created clone replies, "I feel the same as you ... but something here is different. It is a different galaxy according the calculation on the universal map." They look at the visual scenario all around them of the galaxy and the white planet. They are intrigued with the sight.

The ship carries millions of clones. They seem busy maintaining the ship in order. A section of the ship contains a large elongated tube. Inside the tube are many small tubes, which are used for cloning, similar to fetal-incubation. Inside, the bodies are maintained by a pure fluid. The liquid is not water, but a synthetic blend made by them as an artificial support fluid. They need to find planets of living bodies to extract blood for the continuation of the cloning process. They are angry with humans, who created them by mixing DNA from all kind of races and beasts. Their genetic composition is fragmented and confused. The calculation of how they were made is in the past. It was made by Earth scientists. One beneficial outcome for them was that their brains developed faster than those of humans did. They became superior to humans. The ship is their home, their life. They cannot live outside of it.

Me and the other clones note that the probe did not find anything of importance, just clouds upon clouds, white matter.

"Here is nothing we need. There is nothing of importance in this mass of gas. This planet is of no use. Away from here! This galaxy is vast, with trillions of stars and planets. We will find what we need for survival!" Me exclaims furiously.

They depart, leaving behind the clouded gas planet. With a great explosion, the gas planet disintegrates into millions of pieces of debris, ranging from large chunks to the size of an atom.

At light speed, the spaceship disappears in to the dark universe through an asteroid field.

NEW PLANETS

The Others screen the location and find that the asteroids carry some of the minerals they need.

"We are in a good state! This asteroid field is rich in minerals. Engage the z-rays to find all that is hidden in it!" Me tells the other clones. There are twenty of them controlling the comings and goings of the mother ship and the needs of all the clones in it.

The ship flies through the asteroids at a high velocity, extracting all minerals from them by using a z-ray, a precision electromagnetic device similar to a metal detector.

Millions of tons of minerals are extracted from the asteroids. High-powered electrical energy creates heat that melts and separates the materials, then stores them in different compartments, channelling them to their destination by a series of tubes. They also in high heat melt the asteroids containing all iron to build and expand the mother ship and the thousands of inner ships.

They have minerals now, but they need blood, new blood, for the growth and continued existence of their species.

"Is the reserve of human blood plentiful?" Me questions

the controllers.

"For now it is," answers one of the controllers.

"How many humans are in the incubation tubes?" Me continues.

"Millions, with trillions of cells," a controller replies.

They have a department of incubation in the mother ship just to cultivate human bodies for their needs. There are millions of transparent tubes where the Others have humans maintained alive, like a garden producing supplies as needed. These humans were taken before the end of world and the human race. Now the Others are light years away from the Milky Way and the planet Earth.

"We shall employ scout ships to find new planets," Me orders.

"A great plan, Me! When is the right time for this?" the controller asks.

"Soon! We will start the same way we did with Earth," Me replies.

"We need to check the map of the universe and organize a flight," one of the controllers who has already made the calculations says.

Another controller replies, "With planet Earth, we had a holographic map and a guide to the Earth's atmosphere."

"I am aware of this. We will send one thousand ships," Me orders with authority.

They are all in accord, and they send a flight of one thousand large ships. Within it are thousands of smaller ships and thousand of humanoid clones, a special military group. They will scout the universe for new living planets and slaves.

Far away from the Others, in the Milky Way where the Strangers reside, a new planet, the green planet, now occupies the former location of the blue planet Earth. Primo and his brothers arrive home with good news. They are glad to be back after a long and distant voyage.

"Our home is magnificent from up here," Primo exclaims.

"It is beautiful and tranquil," the brothers reply.

They descend slowly without the protective ball. Many of the other clones are flying towards them, welcoming them back home. On the ground with the soft grass, they walk into a large ball that seems to be alive. All the clones are inside to hear the news and the report from the long voyage. C=2 and C=3 fly away, up over the atmosphere of the green planet. Inside the ball, Primo is entertaining the other clones with stories of their encounters in the dark space of the universe.

"Brothers and sisters, enjoy the spectacle of our universe and the grand view of its planets and stars," Primo exclaims with a smile on his face.

Telepathically, they question him. "Did you find the dark energy?"

"No. We did not hear it or feel it. It was just the regular sounds and energy of the universe. But we will stay vigilant," Primo answers back.

They all admire the vision of far space from what Primo, C=2, and C=3 saw.

"We will be vigilant at all times from now on," Primo explains.

"It is our duty to be vigilant and alert," they all reply

telepathically.

"We will guard our planet and galaxy at any cost," he continues.

They all agree with the wise words of Primo.

"As soon you feel any energy—good or bad—you shall transmit to one another."

With these instructions and direction from Primo, they depart, flying in all directions over the green planet as guardians.

In the meantime, the Others have sent small ships out in all directions to scout new planets for the possibility of life. After a long search, one of the ships arrives at a planet with green and yellow clouds. They see a thick atmosphere. Their ship is 15 kilometers in diameter. It is in the shape of gears. The ship's optical infrared camera's thermal image system uses the advanced z-ray and cameras with highly sensitive lenses and microphones. They can sense and detect hot and cold objects.

They find signs of some kind of irregular life form. They send small ships to investigate. From the large ship, they see the small ships going from one corner of the planet to the other, covering the entire surface.

The planet is illuminated by two stars. Observing the large screen, the Others see extraordinary creatures that resemble snails without shell casings, with two large eyes and four antennas.

"A strange creature," the ship's controller says.

"What are they?" a second controller questions.

"They are giant snails, similar to the small Earth snails," a third controller replies, confused.

They confer with one another on what to do.

"Descending is the best way to gain knowledge of them," the head controller says.

"There is no need to send probes," the other two agree.

They send messages to the small ships to return to the large ship, as the large ship enters the planet's atmosphere. They use the same method they have use on all planets in the past, include planet Earth.

They hover about three kilometers above the surface of the planet. They descend to 300 meters to examine the soil. They see it has a different nature; it is neither solid nor liquid. They get out of the ship to look at the creatures and the ground. They wear masks like goggles, and they hover over the ground. They see the giant slimy creatures measuring 50 meters in length and 10 meters in width. There are many of them. The ground is like a creamy, bubbling ocean with a smell of sulfur and strange gas. The planet is the same size as Jupiter. They know what to do. They embark the ship and depart from there at the speed of light, leaving behind a whirlpool of clouds.

"Let us go back and send a scout ship to extract some samples from this place and from the creatures and bring them to the mother ship," the controller suggests to the others.

THE FINDING

After a year of traveling and searching for new planets, the ships return to the mother ship with information. Inside the mother ship, all the controllers of the exploration are in a large room, 50 meter square, in a chamber that looks like a pyramid. It is made of a special material, which appears to be metal from melted asteroids formed into one piece. The controllers are waiting for Me to arrive. They all stand, waiting.

Me arrives with his twenty main officers for the grand report of the expedition. The floor moves and a large table rises. Seats move forward to all the officers and controllers. A larger seat rises for Me. They all sit. The walls of the pyramid change to a screen. In the middle of the table is a holograph of planets in three dimensions, moving outward like a laser beam at the four walls, demonstrating the travels and findings of the exploration. One-by-one, they indicate the new planets discovered.

The first to speak is Me-88. "At this location, Me, we found this strange planet. It is yellow and green with nine moons. Two stars illuminate the planet in full."

"It seems a large planet, Me-88," Me exclaims.

"It is very large … the size of the planet Jupiter … from Earth's star system," Me-88 replies with enthusiasm.

"I see clouds of gas. Is that right?" Me asks.

"Right. It has gas and sulfur." The projection changes to shows the snail-like creatures and the creamy, liquid floor.

"It has monsters?" Me asks, with his eyes fixed on the screen.

"Yes, Me. They only drag themselves on the creamy floor," Me-88 explains.

The creature is shown from all angles, demonstrating the size of the snail and how it moves.

"Samples were taken from the creature and the liquid."

"There is no one alive?" Me asks with a grin.

Me-89 stands after Me-88 sits. It is his turn. He changes the holograph to show the zone of exploration.

"In this zone, about two light years from mother ship, we encountered a red planet. Our probes scouted the entire planet. They concluded there was no life on the surface, but we did detect some life inside, near the core."

Me-89 continues with his new planet. They are looking at the walls of the pyramid. They see the large red planet, with no life and no air or clouds. They also notice tall mountains and deep valleys. It orbits a red sun along with thousands of other planets, moons, and asteroids, all of which orbit in different planes and in different directions.

"As you can see, the images are blurry," Me-89 says, referring to images of what seems to be life on the red planet.

"Do we need to explore the red planet further?" Me questions Me-89.

"Not for now. Possibly in a far distant future."

The commanders of each of the exploration ships show their findings one at a time. Some of them come back with nothing. They carefully measure the universe mathematically. They find themselves in a large zone, light years from the origin of their home in the Milky Way.

The last of the commanders, Me-100, brings forth the findings from his voyage. He stands as Me-99 sits.

"We were closing in on a white planet. It seemed to be a ball of gas. The energy felt strong even from afar. We sent probes at first, but they came back with nothing. The planet was large," Me-100 reports.

"Finally, we got as close as possible. The planets gravitational force was very strong. Our ship remained at a distance of two million kilometers. I ordered fifty ships to depart to scout the planet at close range."

"Is this planet solid in form?" Me questions him.

"We do not know. The planet has an incredibly strong magnetic field. It is impossible to penetrate the atmosphere," Me-100 replies.

"Do you have any hard data for us?" Me wants more information from this white planet, along with visual evidence.

"We have the data from the probes going in and out of the planet's atmosphere, and the ships coming back." Me-100 shows the visual on the screen. "You can see it is covered in white clouds."

One of the controllers says, "We see the probes and ships being propelled with force. Why?"

"That we do not know," Me-100 answers.

"We will go to this planet in the near future," Me responds with aggressiveness. He is firm in his order.

"Now we know of these planets with possible life. We will explore more of them. The galaxy that they are in is large. It has a mass 100 times larger than the Milky Way."

After reviewing the planets, a large holograph in the center of the table shows the galaxy that has those particular planets. They see the mother ship at the edge of the massive galaxy. The mother ship looks like a small moon, a pin in the midst of giant. They see planets the size of Jupiter, Uranus, and Saturn, and planets ten to twenty times their size. The holograph shows red, yellow, blue, gray, green, purple, and colors beyond their knowledge that they have never seen before. There are trillions of stars and planets, asteroids, and gas clouds never known before in their travels. They make a new concept of measurements for the galaxy and the planets they see using the binary system of calculation.

At the green planet, all is tranquil and peaceful. Primo is flying high above the planet, about 300,000 kilometers, vigilant, making sure that all is well. The Milky Way, the sun, and the moon are in working order. He looks down in admiration of the accomplishments of the green planet, home to him and 144 million clones that comprise his family.

Slowly, he descends, observing the planet. Flying into the clear ocean, into the depths, checking and communicating with all the fish, making sure all is right and good. After all the waters are checked, he is satisfied. Then he flies up into the clear sky with all kinds of birds. He also communicates with them. All is good.

"It is a tranquil and marvelous home. I will do whatever it takes to keep it this way," Primo expresses to himself. He knows how the planet was consumed by fire, destroyed by selfish men, engulfed by self-proclamation gods and creators deceiving one another with lies. He and all the clones adopted the dead planet Earth and created a new Earth called green. They pledged an oath; they will protect the planet with their being to the end.

Primo knows that the dark energy will come sooner or later. It is in the universe somewhere. He needs to be vigilant and observant at all times. He calls C=2 and C=3 for an outlook on the Milky Way.

"Brothers, I will travel into the past. The time would be 1990. My cerebrum gives me information to go and investigate 1990."

"If you must! We will be on guard and wait for your return, brother."

THE PAST: 1990

Primo departs into the past at the speed of light, disappearing in the blink of eye into the dark universe, to the year 1990.

Primo arrives in Russia. He observes Russian scientists engaged in sending a spaceship into the Earth's atmosphere, with 660 female and 330 male cosmonauts to experiment in human, animal, and plant DNA before the Americans and the European Union send their own ships for the same study, the splitting of cells and in vitro tests. The Russians discover the knowledge of the speed of sound ten times faster and more powerful than the planes from the past.

They claim to the world that it was just for experiments over the Earth's atmosphere. Five years later, the Russian spaceship gets lost in space, disappearing in to the dark universe. No communication is received from their ship.

Primo travels all around the Earth in that time period. He notices the Russian spaceship traveling at a high velocity. Primo does not intervene in the problems that the humans are causing to the world and to themselves.

"It all begins here! This is the vessel that started it all

for the future of mankind and our beginning."

Primo knows the first problem. He flies 3000 years into the future to the green planet, passing around the Orion Nebula and back. Primo spent five years in the past, but he returns to the green planet only fifteen minutes after he left.

The Others travel at speed of light through the new galaxy, which they named Gears-Z1 (Gears Zone One). They go from one point to another, covering a great amount of space to find any planet with life. When their mother ship intercepts a planet, they scan for life there. The planet is light blue in color.

"That planet is similar to the planet Earth from the Milky Way galaxy ... the origin of our DNA," Me says to his clones.

In front of a large screen are twenty clones, watching as they get closer and closer to the light blue planet. At the same time, they are calculating distance and mass. One of the clones concludes his calculations of the planet's distance.

"We are two light years from the planet."

"At this velocity, we shall arrive soon," Me says.

A few seconds later, they encounter an asteroid belt. Rocks of all sizes, from small to extremely large, hit the mother ship's shields with force. They are half way to the blue planet. The mother ship is protected with an electric field. The electric field is also used to destroy intruders. It pulverizes the asteroids. They draw the debris into storage chambers.

The mother ship arrives at a comfortable distance

from the blue planet. They stop, hovering above the planet at 320,000 kilometers. They observe the planet from the mother ship. The planet is enormous, twice the size of Saturn. It has twelve moons. It orbits a bright star, which illuminates the planet. From the screen, they do not see any sign of soil or vegetation or even water.

"From here we can see nothing. Send probes down," Me orders the top controllers angrily. He is not happy.

In accord, four probes are sent down to the blue planet. The probes are in the shape of a cube, measuring two meters square. They carry 600 camera scopes on all sides with thermal, z-ray, x-ray, infrared, and ultraviolet sensors for detecting hot and cold temperatures, life, and plants. The probes are instructed to make maps as they travel at a high velocity. After a few minutes, they return to the mother ship with information about their findings.

The cubes enter one-by-one in line into a large chamber. They project their findings, showing all that they saw and recorded onto the large screen. The graphics and pictures are clear in 3D. Me and all the officers, controllers, and commanders are observing. The large screen becomes the planet. They see the planet and the clouds from the probes camera scope. The vision they see is that the planet has only liquid, and the liquid is not H_2O.

"I see no life here! It will be impossible for any creature to live here. This liquid is not water like it was on planet Earth," Me expresses.

"Yes, Me. The liquid is carbonic acid, but we can extract the liquid for our needs," controller Me-2 suggests.

"We can take the components for our use by separating them," Me-6 exclaims. He is an expert in oxygen and carbon.

After they analyze everything and acquire maps from the probes, Me feels that deep down below it is possible there is a mass of some kind that the probes missed. The energy he feels is strong.

"I feel a strange energy about this planet! How about you?" Me wants confirmation from the others.

"I, too, feel some kind of energy," Replies Me-2

"We can send ships down," suggests Me-3 as he looks at the screen.

"Given the planet's size, we shall send ninety ships."

As Me confirms the order to the officers, they send ninety ships, each measuring fifteen kilometers in circumference. They depart the mother ship and head down to the blue. The mother ship remains at 320,000 kilometers out in space. The ships fly in formation. Twenty head north, twenty go south, and twenty each go west and east. Down they go into the deep of the liquid planet. The other ten remain hovering up 20,000 kilometers. Once the eighty ships enter the liquid, it changes to a milky solution, much like a foggy cloud. The ships have no optics to see. No sound could be detected. The ships' instrumentation was disabled. The only communication was telepathic. Communication was weak between the ships.

Unexpectedly they hear and feel a tremor, followed by a massive explosion. All the ships that had entered the liquid are shot out with force unknown. They are flung out of the atmosphere like toys.

The commanders from the other ten ships are confused and shocked. They wonder what occurred to the ships down below.

"What just happened down there?"

"What force is that?" a second controller says to the others.

"Possibly a volcanic explosion … shooting up from the bottom?"

They discuss the matter telepathically. Then all becomes clear and they see on their large screens all eighty ships scatter in space, with no damage to any of them.

"This is a force unknown to us," they say to one another.

"Very strange," one of the commanders says.

They communicate with the other eighty ships to see if all is well after they were ejected from the inner liquid. They reply that all is well. They are just confused by the occurrence.

Commander Me-50 recalls all the ships to the mother ship.

"I do not understand … no one captured any of the action?" Me-50 asks.

They replay the video with confusion.

"We are just as confused as you," all the commanders reply.

"A volcanic force is possible, but we have no evidence of that."

"We have no visual proof … only the sound of the explosion."

Me-50 acknowledges the situation and requests advice

from the others before they depart from the blue planet.

"We should send small ships to investigate the surface at close range before we return to the mother ship."

"That is the plan. We will send 900 armed ships to use force against whatever it is that we encounter," Me-51 responds. All the others agree.

In an instant, 900 small ships from all the large ones fly out and dive into the liquid at high velocity.

From inside the large ships, they observe an incredible scene. Me-50 establishes video contact with the other commanders. "What is happening down below?"

"The planet is boiling," Me-56 responds, as do the others.

The blue planet is in motion with large bubbles. Again, there is a large explosion. They see the 900 ships expelled with great force from the liquid below to the air above. Some collide violently with the large ships, causing explosions. Millions of fragments and body parts from the Others descend and disappear into the blue planet's strange liquid. The surviving small ships return, entering the large ships.

Me-50 is furious about the situation at hand. "To all … prepare a counter attack!"

"At what target, Me-50? We have no target in sight!" the other commanders respond with confusion.

"Engage in delivering air impact at will to all sides of the planet!"

"We understand, Me-50!"

The ships begin to fire their cannons. Tubular pockets of pressurized air are expelled from the center of the gears on the ships' cylinders, generating one million kilograms of force per second, hitting the liquid below with incredible force.

BLUE PLANET

The Others are united for the attack. At once, they all fire the air pockets into the liquid, smashing the surface with force. The liquid planet is attacked from all sides. The assault makes waves as high as one kilometer. The waves move up and down. They continue the attack for thirteen minutes without stopping.

From the ships, Me-50 and the other commanders observe the attacks on the ships' screens. They hope to see some kind of life.

"Cease the attack," Me-50 orders the ships' commanders.

They all stop at the same time. They look at their screens and see that the liquid is transforming. Its motion is changing. Instead of waves, now they see giant ripples everywhere around the planet. They believe that they have caused an earthquake over the entire planet. Me-50 sees the liquid agitating and hears a terrifying roar from the planet.

"What will we do now?" He waits for the other commanders to reply.

"We find ourselves doubting whether there is anything we can do here."

"It is time to go back to the mother ship," they all reply.
"We shall, then," Me-50 confirms.

They depart the blue planet, which has turned to dark gray. They travel at high speed to the mother ship. They do not see the change that has taken place in the planet. One-by-one they enter the mother ship and quickly gather at the main quarters with Me to inform him and the Others of everything that happened at the planet. Inside the large room that holds the pyramid, 645 controllers wait for Me and his twenty closest clones. These are the first clones made after him.

A large hexagon table rises from the pyramid floor and 666 seats come out from the table for them to sit on. Inside the pyramid, the peak illuminates the table like a spotlight. The four walls become a surround screen showing the action that took place at the blue planet. The Others look at the planet with intensity. There is no explanation for what occurred there. From their arrival to the return to the mother ship, they communicated telepathically. Then they see the transformation of the liquid from the blue planet.

"Look there! A transformation!" Me exclaims.

What they see is a large formation. The entire planet has turned into a face.

"It that a face?" Me asks.

"It is ... strange. This is the first time we noticed it," Me-50 replies.

"This image of a face ... it does not appear to be human or us."

"We did not see any image or face due to our exploration

and attack," they all reply confused.

"We must make a complete investigation before we attack from the mother ship," orders Me.

They all begin to take measure of the planet, computing the nuclear cells and the chemistry of the liquid. They survey the area around the planet, including the moons, stardust, and other matter.

Me-50 turns towards Me. "We sent 900 ships down to it and lost 60, plus 600 family cells."

"Were any samples collected from the planet for analysis?" Me asks.

"No, only residue from the liquid," Me-51 responds.

Me looks at the screen, observing the image of the blue planet.

"It is a strange image ... the image of amphibious water life," Me explains to all the Others, placing his hand on his chin.

A clone from the science, medicine, and technical division arrives with a carafe holding about two liters of liquid. He hands it to Me.

"Me, this is some of the liquid residue collected by the ships."

"Very well," he replies.

Me takes the carafe. He looks at it and sees that the liquid is moving and changing different colors.

"This liquid is changing, progressing through all the colors that we know from our human data and other colors that we have learned in our travel over the eons past."

They all observe the strange action, looking at one another confused. They have a strange feeling about the liquid.

"We shall run all tests necessary … immediately!" orders Me.

In the meantime, on the outer walls of the mother ship, drops of the planet's liquid are everywhere. They traveled there on the ships that returned. The drops begin to unite one-by-one, making a covering over a corner of the mother ship. Down below, the blue planet is changing again, turning to a white cloud.

"Me! The planet is changing!" one of the clones informs Me.

"The entire planet is covered in a white cloud!" Me exclaims, looking at the screen. He is puzzled at the change in the planet.

From the mother ship, they observe the planet. They use every means necessary to find something … anything … that will give them a feeling of accomplishment. The Others are in a state of confusion. With ultraviolet cameras, heat sensors, x-rays detectors, and many other types of equipment invented by them, they do not waste any time finding out what is happening at the blue planet. But their efforts are for nothing. The cloud is like a shield. They do not see anything.

"This is unusual. Send the Ultimo Probe," Me orders. He is angry and frustrated, eager to find out what makes this planet alive.

They send the Ultimo Particle Penetrating Probe with incredible force down to the planet. All of them watch the large screen, witnessing the probe getting violently repelled from the blue planet back to the mother ship. The probe is smashed and explodes into thousands of pieces.

The Others observe the destruction of their best probe.

Me freezes. "This planet has nothing … just energy within. It has zero supplies for our needs! The liquid, as far as we tested, is of no service to us! We will leave behind nothing! We will depart now and destroy this nothing!" Me is very angry.

They are preparing to eliminate the planet when one of the scientists enters the room with a report on the analysis of the planet's liquid.

"Me, we have some results from the liquid."

"Bring it quickly! I need to know of this doomed planet!" Me is on the brink of madness, his curiosity and the unknown are disturbing his mind.

The scientist holds a cube made of a transparent material. He places it in the center of the octagonal table. Flashes of strobes lights begin to spread around the room. The liquid in the cube its alive. It is changing color, becoming a spectrum of light. They see the liquid in action in a three-dimensional representation on the screen.

"The liquid is acting like the planet. It is agitated!" Me exclaims. He is disappointed with the results.

Then they see the cube turn white. The movement ceases after it turns white.

"This is nothing! I am done with this planet! Destroy it!" With great anger, he gives the order to destroy the blue planet.

The commanders prepare to engage in its destruction. They observe the planet on the large screen. The room changes. The table, seats, and pyramid disappear into the floor. Control instruments appeared in the center of the room. The room goes dark. The screen on the four walls shows the blue planet wrapped in a white cloud. Many screens appear within the large screen. The small screens show the mother ship moving slowly from the planet.

Me and the twenty clones place their hands on top of the instruments, causing a strong reaction of energy, inducing an electrical wave in the exterior metal bars making a powerful lightning charge shoot onto the blue planet. The force equals one million nuclear bombs. It zaps the planet from all parts of the ship.

The blue planet changes to red, like a brilliant star. The Others observe its destruction on their screens.

They depart at the speed of light.

Light years away, Primo and his family of 144 million are tranquil circling the green planet, admiring their home. The new planet is the most magnificent miracle in the galaxy for life in the universe. Communicating telepathically with one another, they confirm the situation around the planet in space near their home.

"It is all well for now," they conclude simultaneously.

Primo, C=2, and C=3 are also beholding the galaxy, flying through it, making sure that all is well. They return home, descending to the lovely and beautiful green planet.

Primo calls everyone for a gathering to discuss the situation at hand and get feedback from all of them in regard to the dark energy.

"Together we need to feel something that is out there! I know that it will come to destroy us and our planet ... our home," Primo says.

They look at Primo. Empathically, Primo senses fear

from all of them. The unknown makes them defenseless and still.

"Let us together concentrate on the universe to find a black hole, where we can travel into a dark matter to the past, to planet Earth, to a time before we transformed the old Earth to the new Earth, the green planet." Primo enthusiastically speaks to all of them telepathically, making them feel a little more at ease.

"How many of us shall travel the universe for comprehensive coverage?" C=3 asks Primo with concern.

"One thousand will embrace the search," Primo responds with confidence.

YEAR ZERO

At once all the clones concentrate and find a black hole two light years from the Milky Way galaxy, their home.

They all agree with the mission, and so Primo, C=2, C=3, and thousand of spheres depart from the green planet. In the blink of an eye, they find themselves passing through the Milky Way galaxy and into the cosmos. Primo and his two brothers are traveling in a sphere. They enter a black hole, which shoots them billions of kilometers away from their current location and back to the Milky Way galaxy, to a point they had calculated mathematically, the same point in space, but farther back in the past. They arrived at zero year, the beginning of humanity.

The spheres search for the dark energy. They see a family of humans, only four of them. They study and find they are parents and two sons, working on a farm. Primo senses something strange with the area.

"Observe! It is different from the last time we visited!"

"It is ... I do sense some energy now," C=2 and C=3 reply. They do not see anything in the air from above the Earth's atmosphere.

"Let us travel a short time into the past and come back,"

Primo suggests.

In the blink of an eye, they disappear and return to the same time and same place, but in a different scenario. They see the two sons working together at the farm. Then they see one of them grab a jawbone of an animal. They watch as the man with the bone in his hand hits the other one, bludgeoning him to death. They see the man dying, covered in blood, while the other one runs from the scene. He stops, and then looks at the sky, confused, disorientated, and dismayed. Then they see the man slammed down on the ground with incredible force.

Primo feels the energy like a jolt of air, the same feeling he and his brothers been had been feeling for a long time. With incredible speed, they circle the planet and find a dark gray, evil-looking, cubic spaceship three-quarters the size of the Earth's moon. They see it just for an instant, and then the vision of the dark energy is gone.

"We all saw the object," Primo confirms.

At the speed of light, they disappear from year zero into space, time traveling through galaxies, stars, planets, black holes, nebulas, past, and future, to see if they can find this destructive evil force and to know what it is and what it wants. After light years of traveling, the only thing they have found is the dark energy they feel. Traveling the universe into the past and future like guardian angels, they see the change of the Earth from bad to worse.

They arrive at the future present, their home, the green planet. When they arrive, all the other clones are waiting for them. Primo hovers above in the air with C=2 and C=3, checking the planet and all living things to account for any changes that have occurred during his absence from the planet.

Again, they all gather at the large sphere for the report on the search by Primo and the other scouts. Primo collects the information from the scouts. It begins to appear around the sphere like an indoor theater.

"There, in the vast cosmos is a strong, dark energy … stronger than here," Primo says, communicating telepathically with his large family.

"As you observe … you will see a large cube for a very short period of time. That is our nemesis, the dark energy," C=2 explains to them.

They all look at them in anticipation of more information.

"Something else we found are very large ships, dark gray in color and shaped like cylinders. They also have the dark energy," Primo continues.

The cylinders are from the Others' mother ship. The scouts discovered them during their search for the dark energy. They see the ships for a fraction of a second … and then nothing.

"We know that they are around the planet Earth. They have been around for thousands of years. For what purpose, we don't know," C=2 exclaims.

"We know they were here before the destruction of the Earth occurred," Primo explains to the group.

Together they project what needs to be done for their protection and the protection of the green planet.

"I agree, my family. We will take action in the past to protect the planet Earth from the destruction carried out by the dark energy with all our knowledge and strength, beyond ordinary limitation with our telepathic capabilities," Primo declares with encouraging words.

Hundreds of spheres depart the green planet into the past of planet Earth. Their destination—1996.

Far, far away from the green planet, the Others are roaming the universe, looking for new planets to conquer and to find new material for their survival. They come close to a new galaxy. From inside the mother ship, they see many planets and moons.

"A new place! A new galaxy! What are the dimensions of this region of space?" Me questions the stellar cartographer.

"Their measurements are unknown at this time," he replies to Me.

"Find out before we enter," Me is excited by this new galaxy.

After an hour, the cartographer arrives with the map of the new galaxy.

"Me, we have a comprehensive measurement of this new galaxy."

"Show it to us!" Me commands.

They walk to the chamber that will show the map in three dimensions. On the table, the cartographer places a cube measuring 15 × 15 centimeters. A hologram appears.

"The 36,000 brightest stars, plotted in galactic coordinates, showing all the constellations in the sky. There are 693 globular clusters. The galaxy measures 366,000 light years across," the cartographer says, describing the new galaxy.

"We will call it GHH-1, for Gold Honey Hive 1. It is appropriate!" Me gives the name to the newfound galaxy.

Traveling at the speed of light, they enter the newfound galaxy. Slowing down, they begin to scout the new planets, coming across a planet as large as Earth's moon. They hover over the planet. Through the screen, they see that it has no signs of life, so they send a probe to investigate more closely. Inside the planet, they find minerals and metals. Within minutes, they extract as much as they can, sucking the life out of the planet.

"We have no further use for this one. Destroy it!" Me orders.

They shoot rays of lightning and massive air pockets. The planet explodes into small chunks that fly through the new galaxy, making a shower of rocks. The Others depart at speed of light.

At the same time, the Strangers arrived at the Milky Way galaxy. Invisible at light speed, they travel the Earth, searching for the dark energy and its creators. With attention, they all quickly communicate from everywhere. They search and search, but the results are minimal. Going from the present to the past and back, they find nothing but the evils of humans and their way of life, dark with negativity and failure to maintain the planet safe and clean. From the past to the present, they see that all men do is fight wars, killing, destroying, cheating, and lying.

"This is our old home! It must have been a good seed that we came from ... good positive DNA! We shall find it before is too late," Primo says to his brothers.

"I know the good seed is here! But so is the evil seed

that made the dark energy," C=2 replies to Primo.

During their travels back and forth, they collect DNA samples from their bloodline, their ancestry of the years and time on Earth.

They travel again, back to 1996. They see a man sitting under a fig tree asleep. They descend, hovering above the ground. The three of them concentrate, and in one second, they know who the man is. They float over to him.

Primo is the one that speaks to the man. The man is frightened by the appearance of the strange figure. He holds very still while conversing with Primo.

"Do not be afraid. I am a friend," Primo begins

"You speak my language?" the man asks.

After a short conversation, Primo and his brothers take the man into the sphere and depart Earth. Fifteen minutes later, they return to Earth. They leave the man and depart in the blink of an eye.

GALACTIC INFINITY

The Others travel at speed of light in the new galaxy to find a planet with life. Me is studying the new holographic map for indication of new planets and stars; he sees a vast space and time at this new galaxy that it will take thousands of years to explore.

"I have a plan to search GHH-1."

They all listen to what Me has to say in the regard to his plan.

"We will send 3,366 ships to scout and retrieve information so we will know how to cultivate this new galaxy. It is a calculation I came up with based on the size of GHH-1. By sending this number of ships, it will take one year per ship. This will shorten the time necessary to survey this galaxy."

They all walk into the chamber with the holograph. Me points to the areas and notes the routes the ships will travel.

"Even with this number of ships, it will not be sufficient to know the truth about GHH-1," one of the twenty clones expresses.

"I agree, Me-6. We will gain knowledge as we go,

calculating mass and sizes through the holographic input," Me responds.

The main aperture of the mother ship opens and small ships are spawned from the mother ship, like eggs from a fish. These ships are different from the others. They are diamond shaped, charcoal gray in color, each with crew of ten. From another door, a million probes shoot out at the same time in all directions into the galaxy. The mother ship is also surveying the galaxy using advanced scopes that produce three-dimensional views, which are shown on screen constantly until the survey is complete.

They find themselves in a pulsating field, which they see through the screen. Me is on guard for the unknown.

"Be alert! Here is a source of energy that will not help us."

The Others, alerted by the strong pulse, look at the screen. They know that can cause problems with communications.

"We have encountered this before, but not as strong," says Me=2.

"Open the pulse channels to collect the pulsating energy," Me orders the commanders.

The opening of the crate is a tube. The invisible pulses of energy are collected. These pulses will supply the ship with power. The mother ship travels close to the speed of light through the galaxy, searching for life. But after one year, the small ships and probes return with information collected from their assigned search area. The commanders synthesize the data from the ships memory banks and place them in a large black box. They take the box to Me in the main chamber. There the box will display the findings as a three-dimensional hologram for all the commanders and controllers to observe. After they acknowledge that there is not much there, they depart galaxy GHH-1.

At full speed the mother ship flies off. It will travel many light years before they are out into the infinite universe in search of new planets.

They warp the space-time continuum. They travel for 3000 years around the galaxy. It seems to them that is has no end.

"We are in a whirlpool channel, possibly near a black hole. Let us be vigilant … the black hole will push us into the fourth dimension."

Me is super intelligent. Being the first clone, his brain is fully developed. He has become the master of the universe.

Finally, after 3000 years, they are out of GHH-1, flying towards the dark infinity of the cosmos.

No one—not even the Strangers and the Others—knows the beginning or the end of time and space in the universe. Traveling at the speed of light, they know they will never reach the edge of the cosmos. The fact is, aliens have few answers for mankind. When the Others arrived at planet Earth, they formed a race according to their needs. Me hates the humans for what they have done in creating them for human glory and scientific accomplishment.

Humans created aliens after aliens created. The Others arrived at planet Earth. They saw four humans, the first humans on Earth. They saw good and evil. They chose the evil seed for the purpose of men's future evolution, leading to self-destruction. He will teach men to lie, to devise ways to kill their own brothers, and to cultivate plants and animals. They taught the humans science, medicine, mathematics, geography, innovation, engineering, and astronomy. They gave men limited knowledge about the universe and what to name the planets, moons, stars, and galaxies.

After leaving the man named Gianni, the Strangers depart Earth. By the time they return, Gianni is dead. Communication between them ceases. The world is changing, getting worse every day. The problem now is that Primo and his family may never find the dark energy and may not be able to save the green planet and themselves.

More spheres depart from the green planet to unveil the dark energy around the cosmos. Their speed is fast, the speed of light. Twenty-two million spheres travel through the cosmos on a grand mission. They are determined to find the dark energy. One hundred twenty-two million remain home at the green planet for security. With them are C=2 and C=3, controlling the situation.

Ten light years from the Milky Way is a planet similar to Earth, with water, land, and tall mountains. Clouds cover the planet. Orbiting the planet are five moons. The planet orbits a bright yellow-orange star. Above the planet's atmosphere are millions of flying objects similar to Earth's satellites. Entering the planet are flying machines, and its surface boasts advanced cities with tall buildings The planet is called Salvation. It is five times the size of the Earth.

What is this place and who are the habitants? The past will tell.

In 1990, the Russian government sent a spaceship into orbit carrying an important payload and cosmonauts. There were 660 women and 330 men on board. These people were on an important mission. They were professors, scientists, doctors, and inventors. The ship also carried seeds, as well as animal and human DNA.

The ship is very large, with nuclear-powered fusion engines. It was sent into space while the Russian government told the people of the world lies about its purpose.

Five years into its research, something went wrong for the Russia spaceship. It was hit by solar wind. The shock killed communications, and the ship moved out of orbit, propelled into deep space in the Milky Way. They lost contact with Earth and control of the vessel.

The Russian government delivered the message to the world in a news conference, telling everyone that the ship was lost in space and all communications were cut.

The Russian spaceship roamed the universe for 100 years. The cosmonauts all survived the disaster. They lived for a long time and aged. Supplies were in abundance: water, food, and nuclear fusion energy. They could last a long time in space. After all that time in the ship, they had children. The cosmonauts began with a family of 990. It grew to 6,633. The ship became too small to hold all of them.

HUMAN RESSURECTION

The Russian ship traveled faster and faster as it fell into the vacuum of a black hole. The ship was pushed into a different zone. Then they felt a pull. Something was pulling the ship with extreme force. They embraced and cried out, "We are all going to die. This is the end of our voyage."

They screamed in terror, speaking Russian. They looked through the screen. The cameras showed they were about to crash on a planet with patches of green, brown, and white. What they see is a planet that looks like Earth. They entered the planet's atmosphere. The ship began to burn like a meteorite.

It crashed on a solid surface. The cosmonauts were okay, with only minor bruises and cuts thanks to the padded walls. They looked at the only screen that was in working condition and saw the surface of the planet. It was covered in deep snow and a large body of water, a gigantic ocean. They did not see trees or animals.

"Could this planet be Earth? The snow here looks like Antarctica," the commander of the ship expressed himself to the rest of them.

They opened the main door of the ship after the

computers told them the planet's atmosphere was safe to breath. The nitrogen and oxygen levels were regular, similar to Earth. The ship automatically decompressed. The automatic door opened and they got out. Outside was a beautiful, large, yellow-orange sun three times the size of Earth's sun. Circling the planet were five moons. They could see everything clearly.

"It is cold like Russia! We will remain inside the ship for now," the commander exclaimed to all.

The night came and they saw the phenomenal sky light up with brilliant stars, illuminating the dark universe. It was very different from the Milky Way and the Earth's solar system. They could see the moons in different locations. There was no Orion's Belt or Big Dipper. It was a long night.

The next day, the commander was up and about, walking outside the ship, looking at the sky and testing the snow. He detected no difference from the snow on Earth. They all came out to admire the new planet.

"I will take out the emergency ship to scout the area," said the commander.

The commander and four other men flew up to scout the planet. The emergency ship was also powered by nuclear fusion.

They flew over the planet for five hours. All they saw was ice and snow everywhere. They returned to check out the ship completely before they traveled a long distance from where they were.

"We will go back up. We need to map this planet. The map of the universe that we have from Earth has no indication of this galaxy or this planet. We are on our own now. We have a new planet, a new galaxy, and a new life."

They went up and finally found a region without snow that had a comfortable climate. The commander had everyone ferried to the temperate region.

"This is our new home. We will call it Salvation. It saved us from death and destruction," the commander declared with gratitude.

Using what they had from the ship, they built homes, farms, hospitals, nuclear plants, and more. The planet became their home. The Russian population increase by the millions. Innovation and development proceeded at a fast pace.

Ten thousand years after the Russian spaceship crash landed, the planet Salvation is now full of life. They do not speak Russian anymore. Their language has evolved into a mathematical system of communication.

The Others are traveling 200 light years from the planet Salvation into an area where space is distorted, visiting millions of planets and galaxies, but finding no life. The population of clones is increasing while the human farm is decreasing. Inside the mother ship are millions of incubators containing humans.

On the screen, they detect a black hole. It seems to be a large one. They are too far away for precise measurement. They know if they get too close to it the pull will be too strong and they will be sucked in.

"We need to get hurled into a time warp by the black hole," Me exclaims, looking at the screen.

The mother ship is 105,000 light years from the black

hole. At the speed of light, they will experience only about ten hours during the trip. They will travel through the black hole, which will pull the ship into the dark nothingness, like a grain of sand.

Suddenly, the ship stops.

"We have stopped. Take observations of the ship, time, map, and speed factor through the black tunnel," Me commands with incredible excitement. He orders his controllers to check the mother ship. Everyone is busy and active.

The controllers check the cosmic map and time. They also measure speed.

"We find no indication of speed. The map tells us that we are in a different dimensional state. Time is moving faster than we know!"

"Impossible! What else is faster than light speed?" Me questions the controllers, confused. "Does the map indicate where we are?"

"No, Me! We shall calculate the time," Controller 1 replies.

The Others have never traveled through a black hole. This will be a first for space travel, and will incorporate new knowledge in their mind. After many hours of studying with the Advanced Micro Mechanics Electro computer, they find a mathematical number for speed, time, and place: 10001101011101100011111101010100101111 0001010010010100100101.

"This is what our computer reads!" One of the scientists shows the result to Me. The others can't comprehend the calculation.

"I remember now! That is time and space that we have

been traveling since the beginning of our creation ... that is the next dimension," Me responds energetically.

"What about the speed?" the controller wants to know.

"Speed is a different matter!"

"We need knowledge of this new phenomenon," Me replies to all of the controllers with concern.

They do not know about the new speed, but they will find out very soon. Faster than the speed of light is the speed the Strangers are capable of traveling. Primo explained the speed to Gianni. It is impossible for the human mind to understand. Primo has the capability to fly faster than one million miles per second.

The Others are confused by the speed and location. The map does not indicate where they are. Me gathers all the commanders and controllers for input on the situation. At this point, the mother ship is still in the dark space.

"Brothers, we shall research and find our location in the universe."

They have all the screens on. They observe the infinity of the dark cosmos. Using the super telescopic eye they see many galaxies billions of miles apart, they calculate that the next galaxy is far from them; but it is the closest of all galaxies.

"This one is the shortest distance! Here we have an indication of a galaxy with a strong core, bright stars, and trillions of planets and moons." Scientist 1 indicates his findings. They look at the star system in three dimensions, calculating its size with accuracy. They notice that there are four suns.

"This is well known to me! Place the Milky Way galaxy on the other screen!" Me commands the astronomer.

The Milky Way appears on the other screen, with the new galaxy over-lapping. They compare the two on all sides. They see it is the same. Now they know it is the Milky Way.

"We are back home," Me exclaims, disappointed.

"Returning home would be of greater advantage for us," a scientist advises.

"Very well, home it is," Me replies.

And so they depart at the speed of light to the new old galaxy.

HOME

Primo and his family are busy searching for the dark energy, traveling to the past, present, and future for 25,000 years. All is vacant; there is no dark energy to be found. They all return to the future, to the green planet, their home. Primo and his brothers discuss what would be best for the protection of the clones and the planet.

"I will continue to search through the cosmos until I find this malicious and dark energy. You will guard our home. Send me any information the moment that you hear anything and contact me when you need assistance," Primo instructs C=2 and C=3.

"Agreed, brother. We hope you find something. May good be with you," they respond.

They embrace, with arms in the Roman style. Primo disappears in the blink of an eye into the nothing of the dark universe.

In the meantime, the Others arrive at the edge of the Milky Way. They stop, hovering. They investigate the zone. They notice a bright star.

"That is an indication. The map is showing us to move towards that star," says the astronomer, looking at the screen.

"I know we are home again. We will find out if the planet Earth exists," Me states.

They know the stars of the Milky Way. The map indicates where to go and displays the catalogue of trillions of planets, moons, and stars. They enter the galaxy through a colorful nebula. They have seen the same formation in previous visits. Not much has changed. They reach the bright sun and hundreds of planets. They believe they are on the right track. They check all the planets. They are different from the Earth's planetary system.

"This is the other solar system that we passed before we found the Earth's system," Me states.

"We are on the west side of the Milky Way," Me-2 informs them.

"We will stay and research it in more depth," Me says to all.

They go on to name everything that is there, from planets and moons to the bright star. They spend two years in Zone One. Then they depart to another zone of the Milky Way. They carry out the same process there, naming everything on their path. They also investigate for life signs.

Then Me makes the insane decision to depart Zone Two and move to Zone Three. He knows that Zone Three is the Earth's solar system.

"We will not waste more time here! We will depart for Zone Three at once!"

At the speed of light, they fly away to the Earth system to find out if Earth has been destroyed.

At the same time, Primo is diligently speeding from

planet to planet in every galaxy to which he can travel, hoping to find that evil, dark, destructive energy. He will know very soon who they are.

The Others are getting close to Earth's solar system. The computer map indicates that the planets are orbiting differently than what they knew. They stop and hover over a large planet beyond Pluto.

"A new planet? Did we miss this one?" asks Me.

"No, we did not miss it," Me-2 responds.

"What is its name?" asks Me.

"There is no name for this one," the astronomer says.

The newly discovered planet was there when they came the first time. It was ignored. The planet is dark gray in color.

"I name it Death! It is a good name for this one," Me laughs.

They extend their super telescopic eye and see that the Earth is different. It is not destroyed, but alive and well. They look at every planet. They see changes with only three planets: Mercury, Venus, and Earth.

"Send a few ships to investigate all planets in the system."

The ships depart from mother ship to tour the planets and record any changes since their last visit.

The Others gave humans scant knowledge about the planets. Man never knew what the aliens gave them. They named them accordingly.

From the sun, Mercury, Venus, Earth, Mars, Jupiter, Saturn, Uranus, Neptune, and Pluto, plus all the moons near the planets. There are comets and asteroids. They also named stars and nebulas. Man advanced in this category by the knowledge and use of telescopes in an impressive way, but all in limitation. What man knows about the universe is a tiny portion of the vast infinity of space, as little as a grain of sand. The planets are station posts for the Others, who map the universe as they go.

From the planet Death, the Others observe the solar system. They see a new planet in the place of Earth.

"Where the hell is Earth?" questions Me.

"That planet is in the same place that Earth was," replies Me-2.

"How long have we been absent?" Me asks.

"Time was not calculated at that point," responds Me-5.

"It is a long time," the calculator of time replies.

"Could it be that the humans survived?" Me again questions.

"No, there is no sign of humans there," the optics controller answers

"We will move closer! Enlarge the optics for greater visual clarity!"

Me orders the optic controller to activate all sensors of the mother ship. Traveling quickly at sub-light velocity, they check everything and remap the solar way.

The small ships return with information about the planets closer to the sun. They enter the mother ship and gather at the main room. They all distribute the information they collected.

"Did you find the planet destroyed? Were you able to get a better view than we see from here?" Me asks with anticipation, eager for the answer.

"The planets are the same as they were before the destruction of the Earth, but better." The commander of the first ship delivers the information.

"What did you see different?" Me-2 asks.

"They are 34,777 kilometers farther away from the sun than before," the commander elaborates.

"And Earth?" Me asks with impatience.

"It seems to be planet Earth ... but it is very different than the old Earth. But it is in the same position," the commander responds.

"How is it different?" Me wants to know.

"It is green. Covered with silver-blue ozone," the commander replies.

"Probes were sent?" Me asks again.

"We investigated with optics and saw nothing."

"Did you enter the planet's atmosphere?"

"No, orders to that effect were not received from control," the commander replies. "We stayed at a distance ... observing the new Earth."

Primo flies rapidly from galaxy to galaxy and planet to planet, through black holes, investigating all he can. When he senses the need of return home, he get a message from C=2.

"Come home, Primo! The dark energy is here, attacking our planet!"

Primo takes off at high speed, faster than he knew was possible. He arrives at his home world. Chaos is upon the green planet.

From afar, he feels the dark energy. It is extremely

strong. For a moment he stops. Observing the planet from the other side of the sun, he sees the mother ship and thousands of small ships attacking the planet, bombarding the planet with lightning and pockets of air, shocking the planet with extreme power. Primo knows it could be the end of them and the green planet. He also knows that they have protective shields on their bodies and around the planet.

Primo communicates with his two brothers in regard to the situation on the planet, so he can make a cooperative plan for the protection of the planet and the elimination of the enemy, the dark energy. The Strangers are not violent. They do not destroy or engage in wars. They are a peaceful creation, the best that men can be. They are all good.

Primo is the first good clone. His offspring are just the same as he. They do not know how to fight or destroy. Now they are all waiting for Primo to command them.

In an instant, Primo is home with the 144 million clones in his family. Immediately, he coordinates a plan to save their home. Two-by-two they depart in the blink of an eye, above and around the planet.

BATTLE: GOOD VERSUS EVIL

The Strangers are flying everywhere in spheres, confusing the enemy in every direction. Lightning and explosions are all around the planet and the ship. High intensity bolts of light strike the planet's protective sphere, but bounce back and strike the ships, causing large fires in space above the planet. The same with the air pockets—all return to the Others, hitting them with extreme violence, sending them far into space. The Strangers are flying in every direction, confusing the Others. The strong bolts of lightning are absorbed by the planet. The mother ship is powerful and can create enough energy to destroy many planets, but they have a major problem with the green planet. The bolts and the air pockets are not doing any damage. They are also unable to hit the flying spheres, which are too fast for the Others' electromagnetic equipment.

"Who are they?" Me ask the second in command of the mother ship.

"We have no visual contact to know who they are!" the commander responds.

"More power!" orders Me.

The small ships continue to bombard the green planet. The mother ship moves at the speed of light towards the sun. They arrive at the corona of the sun and stop, hovering above it.

"Engage to extract energy from the sun at once!" orders Me.

The mother ship begins to extract solar flares. Primo observes the action and flies to them, circling the mother ship at terrifying speed, moving the mother ship away from the sun.

"What is that?" asks Me, shocked.

"We do not know, Me," the commander replies.

"Is the sun moving the ship?" Me asks, confused.

"No, this is stronger than the sun," replies Me-2

"The optics will show us this power!"

They point their optics in every direction, but they see nothing. The mother ship is moving away from the sun and the solar system at high velocity, with some of its external lightning rods damaged. The Others have no clue as to what just occurred. The ship slows down at 33,930,000 miles from the green planet, hovering. The small ships continue battling, but with no effect. Some of them are hit with their own ammunition. Everything they throw at the planet is bounced back to them with force.

"This is a very strange planet. Is it possible we have entered an ultra zone?"

"Perhaps a clone of Earth's solar system."

The commanders communicate with each other in the regard to the new Earth and how it is different from the prior Earth.

They continue shooting at the planet, but the results are minimal. All of sudden, the light silver-blue metallic spheres burst out from the planet up to the Others' ships, surrounding the ships and circling them at high speed, entangling and immobilizing them. They are brought to a halt for a split second, and then the ships are thrown into the dark space, far away from the green planet. The spheres return, hovering over the planet as guardians.

"Retreat to the mother ship!" Me commands all the ships.

From a million miles away, Primo is observing the action intently. He feels that something will occur, so he warns all the other clones hovering above the planet.

"Be alert! I perceived communication from the dark energy. They will re-engage with different tactics towards the planet."

All the Others' ships are inside the mother ship. It closes the hatches, and they move slowly away from the solar zone. They are near the planet Dopa, billions of miles away from the green planet.

"Launch the nuclear cubes," orders Me.

A large square door opens at the bottom of the mother ship. Millions of cubes three meters across shoot out at high velocity towards the green planet.

Like a reverse shower of meteorites, millions of spheres depart the planet to intercept the cubes, redirecting them towards the mother ship. Some of the cubes explode before they get to the ship, others explode at the mother ship, and some just fly by the ship, disappearing into the cosmos. Primo flies to the mother ship in the blink of an eye, circling the ship at an unknown velocity, propelling the mother ship into the infinite darkness of space trillions of miles from the green planet.

"All is well, family! Return to the planet. The battle is over," Primo assures his family.

"Do you need us?" C=2 asks Primo.

"I will be home shortly," Primo replies.

At the same moment inside the mother ship, Me is simultaneously furious and confused at what has happened.

"Never! Never! It is impossible! We are the force of the universe!"

All of the commanders are tense and confused like Me. They do not know what to make of the situation and the lost battle.

"Check all optical and sound transmissions! Find out what we are dealing with!" Me orders the commanders.

They put all their detection equipment into action simultaneously to uncover the enemy from the green planet. They look, but they see nothing, just the planet and the bright silver-blue metallic orb. Passing the planet Dopa, they fire their most powerful arsenal of weapons. Electrical shock and blasts of air burst onto the planet. Dopa begins to burn and crack. Then it explodes, shattering into millions of fragments.

"All the equipment and weapons work well on this planet! Why, then, could we not penetrate the green planet?" Me asks himself and the others.

"We have no answer for that," they reply. "But we shall find out immediately!"

"Reviewing the battle, it is possible we will find

answers," Me-2 confirms.

The Others get busy observing all the options at hand to discover the truth of the situation.

Primo sees all that the dark energy is capable of doing and the destructive kind of creation they are. He follows them for a million miles and he enters the mother ship. Primo moves from one side of the ship to the other, visiting every compartment and place, but remaining invisible.

"This is impossible but true," he exclaims to himself.

He notes millions of incubation tubes suspended and lined up, one after the other, with humans inside. The facility is as large as a city. Tubes and pumps support the bodies, keeping them alive. He moves to another section of the ship, and he sees another incubation area. These are in a pasture. It looks like a clinic of babies. He recognizes that they are clones ... little monsters.

"They are being manufactured in multitudes. They are also clones, like me and my family." Primo is puzzled.

He departs from that dark field to a vast area. He sees large ships and enormous machines. All are charcoal in color. Some are floating; some are still. There are thousands of clones, the Others, maintaining the mechanical needs of the crafts. He flies from there to another area of the ship.

"What is that?" he says to himself.

He notes a large cylinder. Inside, the cylinder is bursting with energy. Primo enters the cylinder and travels inside it from end to end. He discovers that the energy is stronger than the sun.

"This is the life source of the ship!" he says to himself.

THE SHIP

Primo has the power to destroy the ship, but he is good. He does not destroy, but build. He continues to investigate, entering a large area. There he sees a large chemical plant producing oxygen and carbon. It looks like something from Earth with plants, water, and all kinds of living things, with an atmosphere and soil.

Primo goes through the mother ship's compartments in the blink of an eye, observing who they are, what they want, and what their purpose is. He also sees how large the ship is and its contents. He learns how old they are. Primo enters the main chamber, the room where the Others are. He sees Me, plus his twenty clones, the controllers, and the commanders in front of the large three-dimensional screen, observing the universe and watching the planet Dopa being destroyed. The mother ship is flying at the speed of light from the solar system and the Milky Way.

Observing the aliens and the Others, Primo notes what kind they are.

"This is a combination of first civilizations of Earth," he says to himself.

Primo understands their language and can read their

minds. Primo also knows the strange outfits they wear. They look like Mayans, Aztec, and Egyptians in one brown suit.

"This is particular and strange at the same time," Primo thinks.

The Others are checking the screen to see if it has an answer for the battle and the situation. Suddenly, Me looks up as if he sees Primo. Primo is hovering above the large square room where the Others are.

"I feel a strange energy here!"

"I feel the same as you, Me," says Me-2.

After Me announced his feeling about the energy, they all feel it. They stare in the direction that Primo is hovering, but they do not see anything.

"I feel it stronger all over the room!" Me exclaims.

"It is here! I feel the same energy I felt long ago, when we were traveling back and forth from planet Earth," Me-3 replies. They all agree.

Primo takes in all he needs from the investigation. Invisibly he flies away to the outside of the mother ship. He spins around the ship at a high level of speed, causing heat to the point of melting the rods of the ship, fusing them together, so they will not be used again for destruction. Primo begins to propel the ship towards the dark cosmos, far away from the Milky Way. Faster than the speed of light, the ship is ejected into infinity.

The mother ship flies past trillions of stars, through black holes and galaxies in seconds. Primo follows the ship to the end, where he knows it will be in a secure place for the Others. He slows the ship down to their speed and makes sure the Others will never return. His family and planet are safe from the dark energy. Primo leaves the Others and departs from there to his home, traveling one quadrillion light years. The mother ship arrives at a galaxy 100 times bigger than the Milky Way, near the core.

"What happened to us and our ship?" Me questions the others.

"I am numb and confused," Me-2 replies.

The others are also confused, shocked, and dismayed. They look around the room at the screens. They do not know where they are.

"It seems the same. Access the optics," Me requests.

"The optics are disconnected," the controller says.

"No power?" asks Me.

"They have power, but they are out of focus."

"Something in wrong out on the ship," Controller 2 says to the others controllers. He has the mechanical knowledge.

"Let us send the robots to investigate," Me replies. They have smart robots, like human machines.

They send hundreds of robots out into space to check the ship. The robots carry cameras on their chests. They look like eyes. After few minutes, the Others note the most important items on the ship are damaged.

"Where are the lightning power rods?" asks Me.

They observe the screen and see nothing. There are no rods. Then one of the commanders turns and point at the screen.

"Look! Bring the robot back!" Commander 1 orders the robot controller to show the area again.

"The rods are gone ... melted!"

"What kind of force could accomplish this?" The commander is puzzled by what he sees.

After many hours, they finally have the cameras working again. From inside, they look at the screens and know that they are in a different zone of the universe. With no maps to direct their ship and many of the probes damaged, they are in a stalemate with the large, unknown galaxy.

"The strange energy must have caused all this!"

"My feeling of the strange energy indicated it was more powerful than we are, Me." Commander 1, Me-2, and Me-3 are conversing with Me.

"I feel that it is here with us ... invisible possibly," exclaims Me.

They look around the room. Then the room becomes dark, with all the power out. For an instant, they see a sphere in the room. In a flash, it disappears.

"What was that?" exclaims Me.

"A sphere. A ball. The same color that we battled at planet Earth," replies Me-2.

"With us inside the craft?" asks Me, worried.

The lights come back on right after the sphere disappears from the room. The Others look at each other, concerned by the strange force and the unknown power.

Primo arrives at the Milky Way galaxy and stops over the planet Jupiter. He remains there for a minute, observing with attention before he completes the journey home, just to be sure that all is fine in the planets around the green planet. He departs from Jupiter. At the moon, he hovers for a second and contacts his family at the green planet, their home.

"I am back," he calls to the family.

The brothers C=2 and C=3 rise up to meet Primo in the sky. Primo descends. Slowly the sphere covers him, wrapping around and becoming a part of Primo's body.

"I have news to be shared with the family," Primo exclaims to his brothers.

"About the dark energy?" the two of them ask.

"Yes, brothers," Primo answers.

Telepathically, Primo calls all the clones, his family, for a gathering so he can disclose the news. The 144 million clones arrive from all over the green planet at the meeting zone around the large sphere. Primo looks in the air and sees all of them flying. They look amazingly beautiful, covered with a light silver-blue protective skin, like angels in flight. They all enter the sphere, penetrating the walls all around. C=2 and C=3 make sure they all are inside before they enter.

Primo takes flight around the green planet, observing all the animals and plants, birds on land and in the sky, and all the aquatic life, to make sure that the planet is the same as it was before the attack. The other two brothers also ascend high over the planet to make sure that it is all right.

"We will go around the planet in a moment ... checking the planet!" Primo sends a message that the planet will be secured.

"We must have sentinels, great brothers," Primo announces.

One minute passes. C=2 and C=3 arrive. They also are

flying with Primo. They are 77 meters from one another.

Primo flies above them. He centers himself with precision over the center of the sphere, the same sphere that Gianni experienced during his voyage in 1996. The sphere size is seven kilometers in circumference, with no entrance.

"Family! Close your eyes! Concentrate with me!" Primo communicates telepathically with all of them.

"Now, open your eyes and observe!" Primo tells them.

THE SPHERE

Primo and the clones are looking at the sphere. The sphere turns black and then commences to shows the universe. They see stars, planets, and colorful clouds of gasses. All is tranquil and peaceful around the new Earth. They observe the sphere with intensity. A moment occurs when they note strange ripples in space. It seems that the stars are moving. Then they see a large square object, dark charcoal in color, with long rods on all corners of the object. It is the craft they battled previously. They also see the planet being attacked. They see the battle taking place and they know that it is the battle they fought. Finally, they see the large craft being catapulted far into space at a high rate of speed.

"I will show you what happened after the battle," Primo tells them telepathically. He continues, informing and showing them images of his work against the Others.

"With great energy I moved the large ship far into the cosmos, away from our home! There I could have more control over the vessel."

Primo, C=2, and C=3 move around the sphere in unison, changing the scene.

"Here! We are away from our solar system." Primo shows image after image detailing the course he took to investigate and uncover who the dark energy is.

"I pushed the craft beyond our galaxy, as far I could … into the next galaxy!"

"Do we know that galaxy?" comes as a question from the mass.

"Yes, family. It is the Quatro galaxy," Primo replies.

"It seems to be a large ship," the mass says.

"Two-thirds the size of our moon," Primo confirms.

Observing the actions of Primo inside the mother ship, they see amazing things they have never seen before.

"What kind of creatures are they?"

"You will be shocked when you see the truth," Primo transmits in reply.

At this point, Primo shows the images from the ship. The clones are intent in shock.

"Here you see a chamber … they cultivate humans," Primo confirms what they see.

"Are they consuming the humans?" they ask.

"They use the humans' blood and cells to create more of them."

Primo continue to show the next horrifying chamber. He is worried about showing them, because of what they are going to witness.

"This chamber has incubators, for cloning."

"For cloning?" they ask.

"Yes, cloning … like us," Primo responds.

"The same as us? Explain, Primo," they all request.

"I will inform you of the truth in a few seconds," Primo

says. Now Primo shows the environmental chamber, giving them images of a substantially different side of the ship and their way of living.

"You see will see a fantastic chamber of life," Primo says.

He takes them on a tour of the ship so they can understand the size of the craft that the Others inhabit. From one corner to another, Primo makes sure they see everything that he observed.

"Are those all of the ships and the creatures?" asks the mass.

"Affirmative," Primo answers.

"How many?"

"Millions of ships, and millions of clones," Primo again affirms.

Looking at the ship and seeing all the strange things, they are confused and discouraged by what they see. They worry that it will come back to destroy them and their home planet.

Primo looks at them. He sees an unhappy and stressed family. He knows that it is temporary. It will pass, but for the moment, he needs to finish the tour.

"My family, please stay calm and relaxed. The end of the tour is near."

Primo takes them to the upper chamber where the Others had gathered so they see with whom they are dealing.

"This is the commander's chamber. They call him Me. They are all the same. They are all clones like us, but different."

The family is looking acutely at what Primo is explaining about these strangers and their features.

"They know every language, just like us. They know mathematics like us. They communicate telepathically as well."

"Are they as fast as us and as strong?" they ask with concern.

"No, they are slower and weaker than us," Primo assures them.

"Those creatures have many powerful ships!" the mass exclaims.

"Yes, they have many crafts … and powerful ones, too," Primo replies.

Then he shows the ship's tubular source of power that he entered and scanned.

"This large cylinder is 3,366 meters in diameter, and it is 96,300 kilometers long."

"What is it used for?" the mass asks.

"It is a nuclear power source for the entire ship," Primo replies.

"They seem to be an advanced race," C=2 states.

"Yes they are. They recycle everything in the mother ship," Primo says.

"Mother ship? Is that what they call the main craft?" C=3 asks.

"They call it the mother ship because it creates life in all forms," Primo explains

They have been informed of Primo's knowledge that was gained from his investigation of the ship. After Primo delivers the information to the mass, the sphere vanishes.

The mass waits together for Primo to continue his evaluation of the situation and advise them on what to do to keep the planet secured and the family protected from the dark energy should it ever come back.

Primo is calm and self-controlled. He delivers to all what would be best way to maintain order and safety, not just for the planet and its living creatures but also for the Milky Way.

"My family, I will depart for a little while to travel to the past, to search time and discover how it all began. I will discover the facts about the creation of the Others at year zero of humanity."

Primo's plan receives approval from all. "Only if you feel that is the right solution," the mass replies.

"Yes, I feel that it is," Primo confirms.

"How long shall you remain in the past?" comes the question from C=2 and C=3.

"My calculations tell me three days of our time, but 14,000 years of human time," Primo tells them the results of his research from beginning to end.

"I shall commence in 2012 and move backwards to year zero of human creation. I will return with my discoveries concerning the Others and our creation."

Primo departs. He vanishes into the cosmos. In an instant, he arrives in the year 2012. Primo begin to investigate. He knows that it is the right time.

2012: BEGINNING

The aliens had not chosen that time for the cloning of humans, for that time was the beginning, and in 2012, all came to be. That was the year that the first human experiments on cloning began, and the first clone did not survive. The scientific and medical teams knew that the lab was not adequate for tests on human cells or experiments on DNA or the development of human cloning. They were not successful in engaging the process of living tissue. So they began using animals with greater effort. In time, they got what they wanted. Primo was saddened by the act of the primitive humans.

In the meantime, the world governments formed a forum of experts in the fields of science: chemistry, engineering, mathematics, mechanics, and medicine. They called it the S.S.D.O.C. (Space Scientific Development Organism Cloning). The human race was constantly in disarray: wars, famine, civic disorder, the accumulation of laws to constrain people's lives from freedom. It became a terrible place to be. By mid-2012, many terrible things had occurred; climate change, the collapse of the economy, unemployment, and high oil prices rocked the global economy. The challenge to make things better became harder and harder. They had no problem spending money in developing new items, but there were very few jobs for the masses. A year of big change came from developing new machines, including greater spaceships. The space shuttle was a thing of the past. With a new spacecraft and the capability to go faster and farther, it was time to establish a new laboratory for the beginning of a successful experiment to create a better human, a super human. The spaceship program was called the Arc Program. It was the size of a football field, rising one hundred feet high with many compartment used for many other experiments. The first spaceship was *Adam*.

"This was our first home!" Primo says on observing the ship.

In 2013, the ship departed Earth, carrying the purest DNA from the best humans in the history of the race. After a successful take off the communication from Earth's space center to *Adam* were crystal clear. They got ready for the second spaceship. It took about six months. This one they called *Eve*, for mother ship. It was larger than *Adam* with different organism, cells, microbes, bacteria, and DNA from plants, bugs, insects, mammals, amphibians, and human. It was spectacular. Now the hopes of the human race are up again beyond belief.

"Things will get better now," the scientists think.

But they do not know the outcome of all this, that it is all for evil, bringing malice to the human race and to planet Earth. On Earth, everything was very good for a short time, but the time of tranquility and happiness ended with a terrorist attack, a bomb detonated in the center of Tel Aviv. The explosion was large, with more than five thousand people dead and thousands injured. Chaos, panic, and destruction were everywhere. The hope for peace was over. For the next days and months, more problems arose, more deaths and new diseases. The bomb was made to kill and wreak havoc, bringing tribulation to Israel, with chemical properties to destroy the country. The poison from the explosion got into the air and the water. New types of bacteria developed that had no cure. Within the Israeli government, panic occurs. They had been surprised by the attack. They claimed that the attack came from Iran. The councilors and the leaders organize an air strike against Iran. But before they attack, they will need the UN's advice, so they do not make a mistake. An investigation takes place to find the ones who carried out this terrorist attack and determine what country was involved. The UN Security Council is now composed of the United States, Canada, Italy, Britain, France, Germany, and Australia. These countries will make the final decision, including when and where to strike.

For the human race, the situation goes from bad to terrible. The problem now is not just man-made disasters with bombs and chemicals, but also natural disasters: tornados, hurricanes, tsunamis, earthquakes, and volcanoes. And on top of it all, the economy is in shambles. The market is in free fall. The world is in turmoil, and countries are against each other—revolutions and civil wars, the take-over of governments, a lack of jobs coupled with the growth of the world population. It is increasing so rapidly that the planet cannot support it.

The leaders of the world's governments need to work fast to develop means of survival, before is too late. Years go by but it is all empty hands. Instead of making new jobs in agriculture, jobs are created to make missiles, spaceships, and bombs. The decision has been made to go out into the universe and explore new galaxies and new planets, to find a place for a better life for mankind. This is nothing but a dream, an illusion of man and his curiosity. He is not capable of bettering himself. His brain was programmed by the aliens. He will not go anywhere. His capabilities become stagnant.

Communication with *Adam* and *Eve* is less and less frequent. They send a spaceship to investigate, a craft called *Reportus 1*. But in mid-air, everything goes wrong. A terrifying explosion kills all thirty astronauts in the blink of an eye. That stops space research for a while.

In the meantime, in space, the experiment of human cloning is taking place. After a long and intensive effort, the first clone is created. By the next year, there are one hundred of them, male and female, equal. They are beautiful, tall, and intelligent. The scientists named the male Primo and the female Prima. They are the first clones created.

"Here we are, the two of us! The first and second," Primo says when he sees himself and Prima.

On Earth, the population is going mad. There is no place to go. Food and water are scarce, and the water and air are becoming contaminated. The people are dirty, hungry, and dismayed.

China is becoming aggressive towards westerners. They have nothing to say but blame.

"All of the troubles of the world are caused by those from the western hemisphere!"

Eventually, the Chinese government deports Europeans and American from China. They return to Europe. People of European descent depart from all over the world and head back to Europe.

"I have seen enough for now. I will time travel many years to the past," Primo says. He dissolves, leaving AD 2012 and traveling back to AD 7.

WARPED TIME

In the AD 7, the Roman Empire ruled the known world. Primo looks at them and recognizes the DNA that made him and his family of clones. They are descendents of Roman blood. He follows their history and lifestyle up to AD 455. By then the Roman Empire is collapsing from within. Major changes take place in Italy and the world.

Primo sees the change of his ancestors in a different way, from the greatest to great and to destruction. And yet Primo does not find any activity by the Others.

"I do not sense any dark energy in this period," Primo decides.

In the blink of an eye, he departs from the year AD 455 and travels to the year AD 1350. What he sees confuses him. He is horrified by what is taking place on planet Earth. Millions of people are dead or dying from a plague, the Black Death or bubonic plague.

The Black Death is thought to have started some place in central Asia. It travelled along the Silk Road and reached Europe by 1345. It was probably carried from the Orient by rats and fleas living on the rats that were regular passengers on merchant ships. In this manner, they spread throughout the Mediterranean and Europe. The Black Death is estimated to have killed 50 to 70 percent of Europe's population. The plague reduced the world population from an estimated 500 million to a number between 300 and 350 million in the 14th century. Primo was astonished by what he discovered at in that time period.

"I cannot help them or change this era and the tribulation in human history," Primo expresses to himself. "I will time warp to the year 1900."

Disappearing in the blink of an eye, he finds himself in 1900 in Europe, the Americas, Asia, and Africa.

In 1900, relations between Germany and the New Roman Empire cooled down, for an end of slavery in the latter was not in sight. Italy proper has abolished slavery, but even after the end of the transatlantic slave trade, there is enough inner-imperial slave trade in Roman Atlantis and North Africa left, as well as the discrete slave trade with its neighbors that had ended only recently.

The Italians start to drill for oil in the deserts of Libya and Algeria and other places to support their automotive industry and all other type of machines.

Primo witnessed the 1900 Olympic Games in the German capital of Dresden. "Even the games are violent," Primo thinks in solitude and sadness.

While the New Romans celebrate their new century, followers set up the first language school, having translated European works for the world.

"I have seen enough of this era." Primo flashes away from 1900 like lightning, moving on to 1951.

In the year 1951, Primo arrives and notes major

changes in the world. Libya has gained independence from Italy. Great floods occur in the Midwestern United States. That year was also the year of something very important for Primo—the birth of Gianni.

"Baby Gianni, I am at the point of one of my searches. Thank you, Gianni!" With a smile, he departs from 1951 and travels to the time when humans began.

It is raining and gray. No sun shines on the Earth. Primo hovers over the area near the continent of Africa. He looks around for human life and sees two people walking in the rain, conferring with each other.

"It is a man and a woman. Are there no other humans?" Primo questions whether he is in the right era only because there is no calendar at that time.

"I cannot determine the time. I will travel back beyond the present time to the past to uncover their creation."

Primo travels into a dark passage like a black hole. He then appears in a light world with a sun shining on the planet. He flies around observing the planet. He recognizes that it is the planet Earth, before humans.

He searches and searches, but there are no human to be found. The only living things are all kinds of creature and plants. He sees lakes, rivers, and blue oceans.

"I shall measure time as it is now."

Primo measure time in detail. The precise time of its creation is only one million years from the first man.

"I will go further into the past and evaluate time and the beginning of this planet."

Primo goes back and forth through time, covering 500 million years. There are no differences. The planet has life.

It is not so different from the future he searched.

"It is all the same. I will return at time of first man and woman."

He sees the man and woman sitting under an olive tree, dressed with leafs. They are sobbing. Primo hovers about 100 meters over them. They look at the sphere, which is bright as the sun but amber in color.

"Forgive us! Forgive us!" they exclaim.

Primo is confused by their act. The sphere changes color to metallic silver-blue and flies away to investigate the planet Earth and the Milky Way. It returns one year into the future to check on them.

He sees the woman in pain, crying out to the man next to her, "Please, help me!"

The man next to the woman does not know what to do to help her. He gives her water. Primo tries to help from a distance, but he cannot do anything. His powers are stopped by some unknown force.

"I do not feel the dark energy here. But a strange force is keeping me from helping the woman!" Primo decides to depart from that zone.

He flies as rapidly as his ability allows him up above the Earth and into the next galaxy, which he names Seconda. He remains there for twenty years of Earth time. When he returns to Earth, he feels the dark energy passing him in the opposite direction. He senses the energy and he sees the mother ship of the Others.

"It is them! They are the cause of the end of the Earth and the humans and all of creation!"

Primo now knows the cause of the end of the world.

He descends to planet Earth to the area that he left twenty years ago of human time. Primo flashes back one and a half seconds and sees two men walking, taking care of a farm. Then he sees one man kill the other.

"The dark energy inflicted chaos and trauma on the people of planet Earth."

After his discovery of the involvement of the Others in the transformation of mankind and the planet Earth, he knows what to do. In the blink of an eye, Primo disappears into the cosmos, chasing the mother ship. Primo enters the ship. The ship is the same one that belongs to the Others. He visits all the compartments, including the head chamber where he sees the Others, Me and all his controllers. He learns their plan, the leaves the ship and heads back to Earth.

He arrives in the year 2012. Primo goes around the planet Earth, searching for major changes in the planet and the human race.

The aliens had not chosen that time for the cloning of humans, for that time was the beginning, and in 2012, all came to be. That was the year that the first human experiments on cloning began, and the first clone did not survive. The scientific and medical teams knew that the lab was not adequate for tests on human cells or experiments on DNA or the development of human cloning. They were not successful in engaging the process of living tissue. So they began using animals with greater effort. In time, they got what they wanted. Primo was saddened by the act of the primitive humans.

In the meantime, the world governments formed a

forum of experts in the fields of science: chemistry, engineering, mathematics, mechanics, and medicine. They called it the S.S.D.O.C. (Space Scientific Development Organism Cloning). The human race was constantly in disarray: wars, famine, civic disorder, the accumulation of laws to constrain people's lives from freedom. It became a terrible place to be. By mid-2012, many terrible things had occurred; climate change, the collapse of the economy, unemployment, and high oil prices rocked the global economy. The challenge to make things better became harder and harder. They had no problem spending money in developing new items, but there were very few jobs for the masses. A year of big change came from developing new machines, including greater spaceships. The space shuttle was a thing of the past. With a new spacecraft and the capability to go faster and farther, it was time to establish a new laboratory for the beginning of a successful experiment to create a better human, a super human. The spaceship program was called the Arc Program. It was the size of a football field, rising one hundred feet high with many compartment used for many other experiments. The first spaceship was *Adam*.

Primo will not change history. He knows that this will be the work of his creation.

"I cannot change this experiment." He feels deeply involved.

He continues, and in a second he finds himself in the year 2013, the year the ship departed from Earth carrying the purest DNA from the best humans in the history of the race. After a successful take off the communication from Earth's space center to *Adam* were crystal clear. They got ready for the second spaceship. It took about six months. This one they called *Eve*, for mother ship. It was larger than *Adam* with different organism, cells, microbes, bacteria, and DNA from plants, bugs, insects, mammals, amphibians, and human. It was spectacular. Now the hopes of the human race are up again beyond belief.

"Things will get better now"

But they do not know the outcome of all this, that it is all for evil, bringing malice to the human race and to planet Earth. On Earth, everything was very good for a short time, but the time of tranquility and happiness ended with a terrorist attack, a bomb detonated in the center of Tel Aviv. The explosion was large, with more than five thousand people dead and thousands injured. Chaos, panic, and destruction were everywhere. The hope for peace was over. For the next days and months, more problems arose, more deaths and new diseases. The bomb was made to kill and wreak havoc.

"I need to change this history by one second of my time, but I don't want to change the consequences that man has to face," he says to himself, committed but sad.

He stops most of the chaotic action from happening that year, and then he travels to the year 2015.

For the human race, the situation goes from bad to terrible. The problem now is not just man-made disasters with bombs and chemicals, but also natural disasters: tornados, hurricanes, tsunamis, earthquakes, and volcanoes. And on top of it all, the economy is in shambles. The market is in free fall. The world is in turmoil, and countries are against each other—revolutions and civil wars, the take-over of governments, a lack of jobs coupled with the growth of the world population. It is increasing so rapidly that the planet cannot support it.

The leaders of the world's governments need to work fast to develop means of survival, before is too late. Years go by but it is all empty hands. Instead of making new jobs in agriculture, jobs are created to make missiles, spaceships, and bombs. The decision has been made to go out into the universe and explore new galaxies and new planets, to find a place for a better life for mankind. This is nothing but a dream, an illusion of man and his curiosity.

Primo sees the change of behavior in humans and his knowledge saddens him. He reflects on the situation, trying to decide what to do.

"If I intervene to make this change, it will be of no purpose to them."

He knows that the world will change. He sees the future of the human race. It contains disaster and destruction.

APOCALYPTIC TIME

Communication with *Adam* and *Eve* is less and less frequent. They send a spaceship to investigate, a craft called *Reportus 1*. But in mid-air, everything goes wrong. A terrifying explosion kills all thirty astronauts in the blink of an eye. That stops space research for a while.

In the meantime, in space, the experiment of human cloning is taking place. After a long and intensive effort, the first clone is created. By the next year, there are one hundred of them, male and female, equal. They are beautiful, tall, and intelligent. The scientists named the male Primo and the female Prima. They are the first clones created.

For the first time, Primo knows he is the first of his creation, and his female clone is the second. From that point, he knows what course he needs to take and where to begin the search.

He travels forward to 2033 and sees the devastation of the Earth. He researches that time period to find what and who caused the destruction of the planet. He notes a change in human life. The Earth's rotational period has changed slightly. Primo then travels to the other planets to examine if there is any change in their orbit around the sun.

"This is a natural change. It is a galactic shift," Primo concludes.

He visits 350 planets that orbit the Earth's sun to ensure his conclusion about the change and the cause of it all.

"I cannot change the course of the universe! If I do, it could inflict great damage on the cosmos."

Primo observes the movement of the Milky Way galaxy and then returns to Earth in the future.

Now America and China they have a conference, an agreement to a talk with the Imperator of Utopiaromana in regard of their power and ability to destroy them, Canada and Australia.

The year is 2018.

President Carl Chiin-Gu calls the emperor.

"Honorable Emperor Luscious, I have some news of great importance. I am suggesting that you and your empire pay careful attention to this information. It will be given to you only once."

The emperor with tranquility and confidence in his empire replies, "Important news? Of what? I clearly said that dialog between our nations was over. Isn't that correct?"

President Carl Chiin-Gu responds angrily, "Emperor, sir, you don't understand the consequences! If we cannot come together in accord, I will not be responsible for the consequences for your empire and the citizens of Utopiaromana."

The moment arrived, and the sign for which the emperor was waiting came. With that, he cut off all communication for good. That was the end of any relationship with the outside world. The turmoil and panic in America got worse. Violence raged. Chaos was pandemic. The government took action, imposing curfews and a new form of laws.

In South America, Asia, India, and the Middle East, terror consumes the populous along with famine and pestilence. The pandemic continues. No one knows what is going on.

"It is the end of society! We are all going to die!" People scream in panic and horror. They shout from the bottom of their lungs, "God is punishing the world and the human race!"

The dead are everywhere, piles and piles of bodies. The smell of burnt flesh from humans and beasts alike fills the air.

In Utopiaromana all is good and peaceful. The emperor calls upon Australia and Canada to organize transportation for all the citizens to get ready to move to Utopiaromana, to their new home.

"We will be one, in one land. We will not make the same mistake that the Romans made. We will stay together."

By the year 2020, Canada and Australia had transported most of their people with the new spacecraft that the engineers of Utopiaromana created. It was fast and large, with a capacity of 12,000 people at one time.

From Australia to the empire, the trip took only one

hour; from Canada, twenty-four minutes.

Utopiaromana is a place of security with an impenetrable electrical wall circling all its lands and oceans, for kilometer upon kilometer, protecting Europe from intruders. There is no way to get in. In space, satellites observed the whole planet.

In the meantime, America is searching for the best archeologist. They want to find new possibilities and signs that they may have missed in the past. So now the president, Carl Chiin-Gu, has access to all the countries with the help of China's prime minister. They have a free passage to Africa and all Asia, ripe for new explorations. The archeologists and astronomers travel towards the southeast of Mexico. They arrive at an unknown island. They know little of the history of the island and its people. They survey and find no place to enter. After they rendezvous, they contact the president to inform him and to ask for a decision on what to do.

President Carl Chiin-Gu replies, "Did you find anything?"

"Yes. We saw a wall around the island, a very tall wall, about 100 meters. The surface is smooth and the cut is perfect."

"Well, that is different. Maybe something will transpire. I will send a helicopter so you can get in and survey the island. You will have sufficient supplies to remain there until you find something that will help our world and our people"

They reside there for a few weeks, searching for clues and signs. They come up empty. They decide to stay there until they find something of any kind.

In the Americas, the pandemic is out of control alongside diseases and natural disasters. The government is losing control of the people. President Carl Chiin-Gu places a curfew on all citizens.

Angry and out-of-control people take the law as a joke. They become aggressive and more violent towards one other. Revolution is on the rise. Killing and starvation wreaks havoc on the land.

The same fate strikes Asia and Africa, with millions of people dying. The world of the human race is coming to an end. There is no answer to the problem. Leaders take extreme measures. They decide to use force against the populous.

In a different way in Utopiaromana, all is well. There are no complications there or in Canada or Australia. The manifestation of tranquility is beyond anyone's imagination. Outsiders have no idea on what is going on inside of Utopiaromana, and the insiders don't know the atrocities of the rest of the world.

Two months pass. On the island, the archeologists come across a square pillar of stone. It looks similar to granite or marble. It is smooth and brilliant, measuring 6 meters high, 6 meters wide, and 6 meters deep. On the top is a hole 66.6 centimeters across. From the top of the pillar to the center of the hole is 2 meters. From the sides to the center of the hole is 3 meters. From the bottom to the center hole is 4 meters.

On the stone were symbols that resembled Egyptian and Mayan writing. The scholars had a hard time translating the stone. They work night and day with no conclusion. They send a message to the president to send everything they can find on Egyptian and Mayan history and language. The president sends all the material he can get his hands on. After they receive them, they get to work on it rapidly. They don't want to waste any time on this amazing discovery. Forty-two days go by. Finally they find something significant. There was no doubt that the translation was correct. It was verified by multiple scholars. As much as they might want it to be otherwise, there was no denying the message on the stone.

This pillar is perfectly symmetrical, calculated from the Universe Planetary. Once the hole is in line with the planets as well as the moon, it will begin the chain reaction that will eliminate planet Earth and its atmosphere. It will be a dramatic evolution with disasters. This is the key of knowledge, to measure time and dimension for the future of humanity and its end. The countdown for the end of humans and all living creatures will commence on day 730,000 from day 0. The chaos, confusion, and multiple natural disasters leading to the end of the Earth and humanity shall commence on day 14,600 at 0 hour.

After they finally read the message slowly, the writing begins to self-destruct, deleting itself with no trace of its text or its origin.

BACK & FORTH

After all he learns from the year 2018, Primo travels back to 2011. There he notes the problems brought about by man. Primo arrives at the beginning of the year, witnessing major tragedies. He remains there from January to December collecting dates and disturbances.

In the month of January, Primo sees the atrocities that take place. He experiences them in a flash. A man in Tunisia sets himself on fire for a cause, sparking an anti-government protest. He witnesses a flood in Brazil that kills over 900 people. In Russia, 37 people are killed and 185 wounded in a bomb at an international airport. In February and March, a major change occurs when the price of oil increases by 20%. A 9.1 magnitude earthquake in Japan and a subsequent tsunami kill 16,000 and leave more than 4,000 missing. In the same year, Primo sees militaries from all over the planet Earth preparing for war. It is a worldwide effort.

From May through December, he finds that the economy and the wars are making the world unstable. Innocent people are being killed. Dictatorships collapse. Other unstoppable natural disasters occur and many people are homeless or dead.

"I cannot change the history of the human race history and the physical characteristics."

Primo again tries to find a solution to help mankind without changing history. He moves on to 2012 to investigate what happens in that year.

In January, he notes that humans are encountering a new problem. They believe that the end of the world will occur. They expect the human race to cease in December 2012. Primo looks at the calculation of the end of the world according the world's experts and scientists, but before he attempts to find facts in that regard, Primo checks the history of 2012 from January 1st to December 1st. The year brings many innovation and new discoveries by the human race, but also many people die. He travels back and forth from the past to the future and stops in the Middle East in Iran. Within a second, he finds the problem, a potential horror for the human race and the destruction of a third of the planet. He sees a nuclear plant and enters it. The month is June. Iran is engaged in making an advanced nuclear bomb with the capability to eradicate Israel from the face of the Earth.

"This could cause major changes for the human race and the planet. I need to intervene this time for the good of the planet." Primo is concern that it will cause a chain reaction of nuclear bombs being unleashed by countries with large arsenals.

"I cannot dismantle the weapon now. I will wait until they initiate the command for activation." Primo has a plan.

On October 26, 2012, the Iranians are worshipping and praying on the first day of a Muslim holiday. At the same time, the government is making decisions and coordinating its attack against Israel.

Primo is hovering over Iran, looking down at the nuclear plant, observing.

A silo opens and a missile departs from it, propelled by fusion power towards the sky.

"It is time to eliminate this murderous and destructive device!"

Primo controls the missile telekinetically, guiding it out of the atmosphere towards the sun. The sun swallows it like a firecracker.

This event will never be recorded by history.

Primo travels to the year 2036. History tells him the end will come to man and Earth. In the blink of an eye, Primo is there in the future.

"Time is progressing towards the end for the humans and the Earth. I will find the end and calculate the time remaining before the planet and all its life are terminated."

Primo learns something is occurring on the planet. It is not caused by humans but by changes within the cosmos itself. Primo leaps in time to the future, to the year 2037.

"The future of mankind is bleak and inevitable." Primo knows the end. Primo will continue to search for an answer for mankind and the planet Earth.

The year is 2038. The time remaining until humanity ceases to exist is 700 days. The planets slowly but in systematic order move to form a straight line with each other for the final countdown to the end of the planet Earth and all living things, and the beginning of a new galaxy.

Primo decides to leap back in time to the beginning of Utopiaromana. He finds that people are secluding themselves from one another. They are using spies and electronic devices with new technology. They find a few willing white men to do the job. They believe it will be good for the country. They promise the president that they will collect all the information needed and come back to report. At a time when a multitude of Americans were immigrating to Canada, President Carl Chiin-Gu saw the opportunity to send his spies with the crowd. There were ten of them.

The spies arrived in Canada and pass the border just because they were whites. No other races were welcome there. That was a change taking place in Canada. The situation was the same for Australia, which received immigrants from South Africa, boats full of white people. In Europe, too, immigration came from the East.

From Canada an aircraft departed to the EU with eight of the spies on board. They arrived in Europe and began searching for clues and information. When they arrived in the EU, they looked around and were stupefied by what they saw. After they went all over the EU, six of them remained there and two departed for Australia. There, conditions were also the same as in the EU. After a few months, one of the spies jumped on board an aircraft to Canada and returned to America to deliver the news and information.

Primo discover a terrible situation. It is difficult for him to do anything to help the human race. He sees every action taking place. He goes back and forth to keep checking on the new world order, the new city, the new country of Utopiaromana and its people.

"I will voyage into the cosmos and hover over planet Earth to find the dark energy, the Others. I feel that this is their doing, to bring about the end of mankind forever and with it the planet Earth." Primo senses the end.

Second after second, traveling at high velocity, he checks the cosmos and dark matter. He knows the force of the dark energy in and around the Earth's moon. Primo observes the stars and see the Others' ship. He hovers over the mother ship, but he is unable to do anything to the mother ship for the moment. If he changes anything, he will alter the future and jeopardize his family and his world.

"I will continue the search, revisiting the past, back to the year 1996 with Gianni."

Primo arrives and hovers over Gianni's home. He sees Gianni and his family. Seconds later from Primo's point of view, he is visiting Gianni and his cousin, meeting with the high-ranking dignitaries of every country.

As Gianni explains to the leaders of the world, Primo hears all of Gianni's explanations about the future of men, but Primo knows that the changes will not come. Mankind's mind is made up and molded towards extinction. Primo continues to undo the dark side of humanity in a calculated manner so that the history of man will not drastically change the direction of the planet and humanity.

"I can make it better for them by giving them signs. Then they will secure a healthy state of life."

In the midst of the conference with Gianni and the leaders of the world, Primo appears from nowhere. Gianni sees him, and a grin of gladness comes to Gianni's face. As Primo floats about a foot off the floor in the back of the large room, the lights flicker off and on because of Primo's energy.

They try to call the UN military, but not one instrument of communication works.

The leaders stare at Primo's large frame, which is covered with the metallic silver-blue protective skin. They remain still and in shock from his presence, for this is the first time they see his authoritative and powerful presence.

"Ladies and gentlemen, as I said before in my holographic appearance, the end for your race is near. You need to change your lifestyle and habits. Get rid of all mechanisms of mass destruction, for the sake of humans and all creation!"

Primo is telling them the truth about their future, which is at hand, as he floats all around the room. Then he shows all religious books from all over planet Earth in every language on a large holograph.

"Here you have many gods, many ways to follow. And because you misunderstand these gods, you unleash wars against one another. This will end your race!"

Primo changes the holograph to a future time on Earth, showing a third world war as he floats towards Gianni.

"Gianni, you did well, but it is finished. They will not listen to you or me. We shall keep contact at all times." With that, Primo disappears.

Primo flies high into the sky at the edge of the Earth's atmosphere and stops. He hovers over the planet and then looks down.

"A beautiful planet is Earth. Men do not realize what he is doing to the planet and himself." So Primo listens to the cries calling for help from all the people below. He hears every single one and all the noises.

"The Bible also says to pray to God for help. Every religion calls to do the same, but I do not see or hear any response from the gods ... not even from the mysterious Jesus. They are alone to the end."

Primo has a plan for the human race without altering history. He departs from that era and mathematically measures time and the physical matter of the universe, including the gasses, planets, and stardust, to unveil the truth and find the footprint of the Others' mother ship.

"The pieces needed to be placed in formation. The calculations will take time. I will find and procure all particles." Primo's concentration is intense.

THE SEARCH: PART I

Primo time warps to 1922, searching that time period for comprehensive knowledge of human history and events. He finds that the year 1922 is peaceful, with very little happening. It is just after World War I and the world is holding together and dealing with the situation.

"There is nothing of major concern. Just small problems are manifested at this time."

With that, Primo departs and jumps to 1945 at the time of World War II. He scans some of the history, and then continues back to the future to 1996, the time when he began with Gianni. He hovers over the planet observing the events and collecting facts, calculating precisely the time for that era.

"I will understand this time. It is what I need to configure the past, present, and future for the good of humanity and the good of my family."

Primo departs from 2012 in the blink of an eye. He travels into the future, to his home, passing and visiting all the planets of the Milky Way. Primo sees many things that have changed. Planets are in a strange configuration in the Milky Way.

"This seems to be a natural configuration of the solar system, a rotation that occurs every seven billion years. They form a galactic line." In seconds, Primo calculates the matter, gasses, and electrical energy of the Milky Way, the home of Earth, the human race, and the green planet, the new Earth. Primo arrives home and gathers his family for a conference. He will report the news to all. They get together in the large sphere where Primo will show all the data he collected from the cosmos, including that pertaining to the Earth's galaxy and a small part of history at the end of the Earth. He uses a digital holographic image that simulates the reality of the events and circumstances of the Earth and all living things.

"Family, what you will see is the consequence of man's choosing, the end of their race and the destruction of planet Earth."

The sphere initiates the pictures. The vision is horrifying to the clones' eyes. They do not understand why humans are so evil and cruel to themselves, and how the clones' DNA could come from the bloodline of man.

"We are in agreement. I will time warp to the past to uncover the history that we do not know," Primo expresses to everyone.

The two brothers, C=2 and C=3, are concerned that Primo will go alone.

"How long you will stay there, Primo?" they ask.

"As long as needed," Primo replies.

As he departs from the future and his home to travel to the time of the Egyptians, he says to his brothers telepathically, "I will return with new knowledge."

Primo arrives in 3000 BC. He notes that the Egyptians are advanced, civilized, and creative, but malicious. He travels back to 10,000 BC to see the first Egyptians, the beginning civilization. There he senses the dark energy.

"So, it begins here" Second by second, he measures space and finds the mother ship of the Others. He remains with them, tracking them and their activity. Primo learns they are the cause of the events on the planet. They are the ones who changed humans, incorporating malice into them.

He travels back to zero time, the beginning of man in the year 12,000 BC. He finds the initiation of Cain by the Others, who inject him with malice, making Cain the source of all evil that man can do. Primo begins to collect data and traces the dark energy to the time of the Incas and the Mayans, through to the Romans and the Americas. He finds a continuing evolution of man for the worse.

"I cannot change the fate of the human race. The end will come to them. I can only eliminate parts of the destructive matter they build." Primo will be the custodian, the guardian of the universe, helping the human race to survive.

Primo travels to 2012, where he remains for a long time investigating and keeping in check the human race's activity to the millisecond, seconds, and minutes to the hour, days, and years to come. But for now, he will stay to understand human race.

"This is the optimum time to have an impact on events from here on." Primo calculates time and space from present to future.

He then transports himself to the end of the year 2012 so he can see for himself the time of the completion of the calendar by the Mayans. He investigates for the entire month of December. He finally has the answer.

"Man has it all wrong about the end and everything he believes. He delivers false information about the future!"

After he concludes that the time for man is short—not because of the Mayan calendar but because of human actions—Primo time warps to the year 3000 BC. He moves forward from there, investigating dates from the Mayan calendar up to 1 BC. He studies their lives and their ideas about the end of the world or infliction of the human race.

"Here is the dark energy." Primo feels the dark energy around the galaxy. He sees the mother ship and the activity of the Others, who are manipulating the Mayan people.

"I do not see the calendar at this time. I shall visit AD 33." Primo arrives in the Americas, where he notes the land of the Mayans has changed, but still finds no clarity with regard to the calendar. Within seconds, he senses something is going to happen to the planet Earth.

"It is strange that I did not feel this before in my travels."

Darkness covers the land for hours, an event which later came to be referred to as the "crucifixion eclipse." He follows his senses to the Middle East, to the land called Judea and the city of Jerusalem. He treats the darkness according to standard reports of eclipses or periods of darkness. Using this period of darkness as a marker, he interprets the event as a solar or lunar eclipse.

Primo has no idea that this event occurs because of an

act of God. It indicates the death of Jesus Christ.

"The dark energy has no involvement with this. It is an unknown phenomenon." For the first time, Primo is confused. "I will remain in this era around AD 33 according to the Roman calendar for a while, collecting data."

He observes the time period and the actions of the Romans and the rest of the world. Primo sees the Romans at the height of their power and notes their accomplishments in the world. He also notes the dark side of the Romans and their capacity for horror and the death of innocent people.

"This race is part of my blood line, yet I cannot comprehend their abominations and their darkness."

Primo leaves that time to explore. The only fact he cannot figure out is how the death of a man can change some people and bring about an earthquake and an eclipse.

The time he investigates next is AD 55. He finds no conclusive data that help him understand. From that time, he warps to AD 77. He chooses odd numbers in a mathematical and calculative way to a precise time in the history of mankind. Primo follows Roman history for ten years from 77 to 87. He learns of a strange occurrence. The power of the mighty Roman Empire is collapsing within from self-destruction and death. Primo investigates and discovers a major devastation in the region south of Rome in a city called Pompeii. In AD 79, the volcano Vesuvius destroys the provincial Roman city of Pompeii. Most citizens of Pompeii have no clue of the volcanic activity that is lurking at their back door.

The prosperous residents of this provincial Roman city of 20,000 plus probably see Vesuvius as just another beautiful mountain and accept the rich soil as a gift from the gods.

Primo hovers over Pompeii a day before the eruption of Vesuvius. He sees the beautiful city and its people in peace and tranquility. In seconds, Primo knows the history of Pompeii. Life there was good for 1000 years, but the end is coming soon.

"If I gave them some signs, it is possible some of them might be saved. But if I tell them that their city will be completely destroyed, they will not accept that reality."

Primo turns his attention to two days before Vesuvius erupts. He begins to produce some smoke from the mountain to warn the citizen of Pompeii. It continues for a few days. The people see the smoke, but they do not pay attention to Primo warnings.

On the morning of August 24, Vesuvius erupts and Pompeii is obliterated along with many other nearby towns. Primo has been generous by giving warnings and signs that disaster was imminent for the city and the people of Pompeii.

"I will stop the poison from traveling further and contain it here. I will save some of them, and history will remain the same as written." Primo saves many Pompeians from the enormous devastation.

"Again, the dark energy is here!" Primo is distracted by the dark energy. After he makes sure that the people are safe, he warps time to the second before the eruption.

"I can feel the dark energy at hand. There they are!" He

sees the mother ship hovering over the planet Earth near the moon. The mother ship remains only a few seconds after the eruption of Vesuvius and then departs. Primo follows them into the future.

THE SEARCH: PART II

Primo will go where no other can go to find truth and to help mankind and the planet Earth. He will save thousands upon thousands of humans from the devastation of the end of the world. He will do what he can to keep history unchanged, but he knows that the end is imminent for the human race.

"How do I convince humans to change their dark habits and save them from destruction?" Primo looks for a new plan of action.

Primo transports himself through history to 1863. He arrives in the middle of the American Civil War.

"Destruction and death are rooted in the human mind."

He sees devastation and bloodshed on an immense scale. He is saddened by the activity of men and their ways.

"I will go to the beginning to find out how this started." Primo knows the history of the Civil War. In seconds, he is in 1861, the early stages of the war. He studies the history of that time period and the leadership of America. He finds the political affairs troubling and chaotic, bringing the American people to the brink of disaster that threatens to destroy the new country.

History tells us that in 1860, the Republican presidential election was won by Abraham Lincoln. He strongly opposed slavery in the whole United State of America. Things were not so good for Lincoln and the Republicans, and right before Lincoln's inaguration, the Confederacy was formed by states where slavery was leagal. President James Buchanan failed to deal with the secession, but Lincoln and his party would not start a civil war, for many good reasons. But that did not matter, and both sides prepared for war. It became one of the bloodiest wars ever.

"I will add this information to my research," Primo says. He will try to discover the facts and the cause of it all. "Here is the problem: the two sides do not want to stop!" Primo is determined to find the quarrel between the two. Primo is sure of that history and he knows that it all began with the Others distorting history for the human race.. Lincoln tries to simmer down the problem, but it was in vain. Neither party agrees on a solution for the good of the country and its people.

"This situation will bring havoc to the future of the human race and the beginning of the end." Primo is concerned.

Even after the end of the war and slavery, the country is in shamble. Primo departs to the the years 1862 and 1863. "Again, this conflict between these two commanders is agressive." To Primo, his findings make little sense. He cannot understand how the generals and commanders like Ulysses S. Grant and William T. Sherman could not come up with solution for the good of the American States. "This bloody war will turn life backwards. It will cause the Earth and man to choose a path with no return."

Primo saves many people from devastation. This is an impossible task for anyone to comprehend.

"I will time warp to 1914, where I will find a new situation in human history." Primo knows that is the time of the first major World War. He knows it will be horrific, but he needs all the information from that era.

When he arrives in the future time, he finds himself in the midst of a more vicious war than the Civil War. It is the future, 1914, and the conflict in the world is escalating. World War I lasts for five years. Millions of people die from weapons never used before. "The human race is vanishing slowly by their own doing even without the help of the Others. They will exterminate themselves."

THE SEARCH: PART III

By its end, the war brought more chaos and destruction than all prior wars.

"This is unconscionable hypocrisy! They use the word 'love,' but I do not understand the human philosophy and the things they do to one another." Primo is again confused by man, the source of his DNA.

The history he uncovers is devastating to him. Again, he saves millions of innocent people through the wars he finds traveling through time. Because he can go from present to the future and to the past quickly if he needs to change small events in human history, he can do so with little change to the overall structure of human history. The recorded history remains the same.

Primo goes to 1945 to research World War II. "I see here it will begin ... devastation with nuclear armaments. The human race will not be the same again" Primo determines that the end of World War I starts a series of events that bring about World War II, which lasts for six hard and bloody years and beyond. This second war hit the world from every direction in a massive way, on the land, sea, and air. "I see that the outcome of this war will be the start of nuclear devastation, which the Others precipitate with knowledge and technology in order to bring about the destruction of the human race.

Primo's investigation has been completed. He travels to 1942 and goes to Eastern Europe. It is the time of the German offensive codenamed "Fall Blau," which marks an escalation of World War II. There, Primo sees that the dark energy surrounding the continent of Europe is strong. Within seconds, he uncovers the mother ship. The Others are in the area. He checks the ship and sees that they are in control of the war, manipulating the action through one particular man, Hitler, who is one of the Others' seed.

Primo goes back to 1939 and finds that Germany started the war to gain power and to take over the world.

"I will destroy the malicious and destructive weapons for the good of the world," he says. "My interference will save millions of innocent people."

Primo acknowledges the evil of Hitler, his manpower, and his ambition in using the most devastating weapons. Primo arrives at a Krupp factory. Hovering over the plant, Primo learns that the Germans are building a new weapon, a large cannon similar to the Bertha cannon, but more powerful and stronger. It is to be used with nuclear projectiles with a range of 1000 kilometers. After he destroys the cannon, Primo departs from 1939 and travels to 1945. There he sees that everything is good and moves on.

"I cannot allow the end of a country and its people with this nuclear devastation. I will stop the largest bomb from been constructed by the Americans by changing history by a microsecond." Primo saves many Japanese lives and Japan as a country.

The American government put together three atomic bombs to destroy all of Japan, nearly 400,000 square kilometers and millions of people. The plan was for the complete eradication of a race. The world knew of Little Boy and Fat Man, but there was one other.

Primo allows the first two atomic bombs to be detonated over Japan so he would not change the history of man and his malicious ways. The larger, more complex bomb was more powerful than all the other bombs put together. They called it X-MT (Exterminator Megaton).

"It will not detonate or be used in any other place on Earth. I will make sure of it. The schematics for constructing other malicious weapons will be destroyed." Primo knows of the nuclear bombs of the future.

With all the information of World War II collected, he leaves 1945 for 1951, arriving at the beginning of the Korean War. "I will observe 1951 for the beginning of this nonsensical conflict. Even the history of this event is evasive."

Primo remains distant from the history he uncovers. He knows that nothing he can do will make men responsible and moral, fit for the care of themselves or the planet.

"I will save some of them, especially the innocent."

In saving many people from the country of Korea, he notes another disturbing act by men. "It is malicious for a man to invoke the truth of his actions to his own kind. It is very disturbing!"

The sequence of events that takes place is unbearable, day after day and month after month. Primo departs from

1953 and travels to 2012. "I know now that integrity will never return. I have no chance of changing them. Humanity is doomed."

The end is certain for planet Earth. Primo, the Stranger who came to save the world and humanity from destruction, who has power beyond imagination, concludes that he needs to take action in different manner.

GOD?

Primo will not change the planet or men. He will not continue to guard the Earth from the Others. Primo knows that the Others departed from Earth orbit and left the Milky Way galaxy.

"It is done. The Others implanted the malicious dark energy in man. There is no going back to revive them."

Primo records all that is in the history of mankind, all the good and evil in every microsecond of human history, straight from Cain. Traveling at high velocity, he time warps through quantum singularities, black holes, nebulas, galaxies, and planets, making the cosmos his backyard in search of a home for the new, meek humans of planet Earth, to save them from destruction and the termination of life on Earth. He wants to find a planet suitable for the new humans.

"Out of seven billion humans alive in the year 2012 and the 11 billion in the year 2035, I calculate that I need 1% of humans, plus DNA from the past and collections of plants and cells from all that the Earth produces. It is needed to form a starting point and the beginning of a new human race. I will also record all the good accomplishments of the human race."

Primo will find a new planet for humans and other living things from the Earth to develop a new species for a perfect life.

"From my search of the cosmos and the indications it gives me, I know the planet for them will be here in the Milky Way." In seconds, he arrives at a planet far, far away from Earth and the solar system. "This will be their new home!"

The planet over which Primo hovers is at the edge of the Milky Way. It has the same appearance and a similar size to Earth. It has comparable water and soil.

"This planet is the one I investigated telepathically during my search. It will be the one just for human habitation." The planet has two moons. Many other planets similar to Earth circle a bright star like the sun.

Because of the Strangers—Primo and his brothers C=2 and C=3—the humans gain knowledge of good, increasing their capability by ten-fold with inventions like the telescope, satellites, and engines to make magnificent ships, as well as advancements in human enhancement to benefit life and the planet Earth. By having this knowledge, man can explore the universe beyond the Earth's atmosphere far into the cosmos, but with limits on how far he can go.

Man discovers many fantastic things and amasses vast information. He learns about planets, the sun, the stars, gasses, meteorites, and moons. He names them according to hints that the Strangers give to man. They learn that the Earth is round, about the moon and its value, and about the sun that gives life to Earth.

"I shall go back to the past, to Gianni and his meeting with the leaders of planet Earth. There is one piece of information that I still do not know."

In a second, Primo finds himself in the midst of the assembly of leaders where Gianni is diligently instructing the group. Then Primo observes the changes taking place at the assembly.

"There I am in the center, communicating with the leaders."

Primo is observing the time when he answered questions, including a particular point when Primo had no answer to a question at that point in his exploration of the history of the human race.

"I recall that question. I lacked that knowledge."

Primo goes back and forth in time by only microseconds. "I will uncover the truth and learn the importance of the question."

The question was from the pope. It was directed to Primo. The pope asked him about God. At the time, Primo did not know how to answer. Now Primo has a new task—to learn about the phenomenon and mystery of God, to add God to his knowledge.

"That is the question. I will begin to research who God is. I will travel back in time. My starting point will be Cain and his journey again."

Primo covers the world's history and the history of man and his advancement. He finds scribes that write in all languages on walls, pillars, and tablets about God or gods. From the beginning to the end of the world, people worship or believe in something or some god.

"Man is confused about gods. The first people believed that the stars were gods. Then from Cain and others, they were controlled by the Others. For them the Others were gods. Then the Sumerians, the Egyptians, and the Mayans arose." From the east of the world to the west, he sees significant transformation in genetics, created by the Others. "I do not see any God or gods." Primo calculates time and concludes that he needs to find the precise moment in history to find the answer.

"History tells me they have BC and AD, the before and after. These are the elements of their history. I will fix a one-to-one correlation mathematically with time and space, the creation of planets and stars."

Every man in history studies the stars, the moon, and the sun for time and calendars. The calendars agree and disagree at the same time.

"The human race … they are off with everything they do!"

Again, Primo will go on a search for the truth of everything that man did from the beginning. His mission will be to find God's history, which man claims.

"I shall have all the writing information that man recorded through his lifetime. I know it contains much false information. Because of that, I need detailed measurements in time and space, in human history as recorded."

He begins to collect all the religious writings material, books from Christians, Muslims, Buddhist, Hindus, and all others that man made for his own gain and search for power.

"This will constitute new knowledge about man that I did not have."

In seconds, Primo goes through the religious and non-religious writings that man recorded in an orderly matter.

"This is very strange and intriguing … here things are more the same than different. They believe in a God of peace and war, love and hate, heaven and hell, and a spirit power received with faith."

Primo pauses in his reading, confused again. "I do not know about these: spirit, faith, love, and peace … or heaven and hell." He continues with his new knowledge to understand what man is trying to find or to accomplish.

His first exploration will be the first man, about whom he reads in the Bible. "I will warp time in microseconds to the beginning, as I read in this text from the part of the Bible they call the Old Testament."

He arrives at the time of Cain's birth, the first baby and second man alive. His father and mother are next to each other.

"He is a beautiful boy isn't he, Adam?" Eve exclaims to Adam.

Primo is witnessing the first human baby boy.

"Well, here we are. This is the first man that begins all the dark history, a man called evil. I will go to time zero of creation to see Adam and his flaw."

Primo finds himself at paradise, where Adam and Eve are walking in the dark in a bad rainstorm; he cannot go beyond that time into the past.

"This is the starting point for my search. I cannot see what occurred before this."

In his travels and search through the millennia, he finds that man is confused in respect to his religious beliefs, holding chaotic and complicated ideas of these teachings from birth to death.

"All religions have a God. The Christians have Jehovah and his son Jesus, with many prophets delivering God's message."

In studying the books of the prophets, Primo is confused by religions like Islam.

"Muslims have strange doctrines and follow the teachings of a simple man called Muhammad. This man went from being a warrior to delivering tribes in the Middle East in the name of what they called 'Allah.' He killed and destroyed villages and many innocent people."

Primo is stupefied at what he uncovers from the writing and the disciplines of this people called Muslims and what they do to follow the teachings of Muhammad, who claimed to be the messenger and the prophet of God—Allah—even killing innocent people for their purpose and the false promises after life.

"Again man is disoriented in their beliefs and guidance by the words they claim are from God."

He comes to the conclusion that all religions are evil and dark. Their teachings are to love and hate, build and destroy, sacrifice, take, and worship anything and everything for one purpose: the intent of self-indulging power and fame.

They speak of peace, kindness, love, helping, giving, and faith with the help of the Holy Spirit. What does that all mean? Where is this Holy Spirit?"

Primo does not understand what the Holy Spirit is, where he comes from, or who he is. He cannot enter that history of faith. Primo does not have a spirit.

He keeps searching countries and races. In the eastern part of the world, he notes that the people there follow the teachings of someone called Buddha. "I do not comprehend this attitude of the human race. They seem to be worshiping this statue of a creature out of the regular health state."

He finds the same things over and over again, with all the gods that people have and worship. They all believe they will make a difference in their lives, from the first tribes of the Israelites through the Egyptians, Greeks, and Romans, until today.

GOD & JESUS

"I am at the end of my search. The only one I will continue to explore is this God Jehovah and his son Jesus. The history of Jesus is short. It says that he lived only 33 years of man's time. He is the descendent of a king named David"

Primo decides to investigate the ancestor first. He travels back in time to the era of David. There he observes that King David has malicious ways. He is a warrior and a killer, and his morals are weak.

"Through human history, everywhere I search I find darkness and chaos. Humans are savage and malicious with each other and all creatures."

Again, Primo is exhausted by his findings. He goes back again into the past to the beginning of man. He sees three people under a tree: a man, a woman, and a baby. He knows that baby is Cain. He travels ahead and notes that the same two people, Adam and Eve, have a second child named Abel.

"I did not understand this before, but now it is clear to me ... the particulars of the human story: the one man, one woman, and the two boys. Here I see the dark energy and the light energy coming from them. It originated here! I will go to the time of Cain, when the Others implanted the dark seed of what man calls evil."

Primo now emerges with new knowledge of the human race, himself, and the Others. He knows now the truth of the ways of man. He needs to find more about this Son of Man, this Son of God, who He is and why he has no records or knowledge of this intriguing history.

"It could be as deceiving as all the information collected from their writings. I shall micro-matter life on the planet Earth before Adam was made into a human life."

Primo warped time to before Adam, searching for the beginning of humanity. He wants to learn if anything existed before Adam and where he came from genetically. He continued the search through millions of years. His conclusion was the same: no humans or strange life. The Earth was the same until Adam. He finds that man has the wrong information with respect to his history.

"It all changed with Adam and Eve!"

He travels back using a mathematical time, observing Adam and Eve to discover their birth and who made them. "The only history of Adam and Eve I can uncover is the two of them walking in the dark rain to nowhere together, and this is the beginning of mankind's history. I do not find his creation or the maker described in the written history that humans have about God."

He hovers over the planet. He sees the only humans: Adam, Eve, and many children. People of working age care for the land and the animals. They are a hard working family. Primo at high speed checks all locations on planet

Earth for other humans. He finds that the tribes are all descendents of Adam and his children, including Cain, the evil seed.

"Adam is the first of all the humans. I cannot find his creator. I will use telepathy and open his memory to know for myself about his creation and his God,"

Primo sees Adam walking alone. He hovers over Adam. "Adam! Adam!" Primo calls his name.

Adam looks around and sees nothing. "Who is calling my name?" Adam replies.

"Do not be afraid. I am a friend."

"What do you want ... friend?"

"Knowledge about your creator," Primo informs Adam with a soft mind message. Then Primo appears to Adam, floating in the air about ten feet above him.

"Are you Lucifer or God?" Adam questions Primo.

"I am neither."

"Who are you then?" a frightened Adam replies.

"I am you, from the future."

"What is this future that I do not know?" Adam says, calmer.

"I will explain using a simulation so you can understand."

Primo descends and slowly his head is uncovered by the protective second skin that covers his whole body. He touches Adam's forehead, and Adam becomes sleepy. In one minute, Primo shows him his lifetime with his family and that he lives for over 900 years, producing many good things from farming to the birth of many tribes, including the population of Earth.

Adam acknowledges the good and evil of mankind

from his own son Cain to the end of the world. Adam wakes up with his eyes fixed on Primo.

"Are you well?" Primo asks Adam.

"I feel compressed with new knowledge. It brings me distress because of all I inflict on the human race and the world." Adam bows his head down to Primo with sadness and tears in his eyes.

"The horror of all this is not of your making. It is a seed planted by the dark energy. Man inherited the core … what you called evil."

"What can I do to change this evil and to make things better for the humans after me?" Adam asks Primo, greatly concerned.

"Nothing, Adam. You cannot change history as it is. Neither can I."

Before Primo departs from his time with Adam, he has one more question to complete his search for God.

"Adam, tell me … who is this God Jehovah?"

"Jehovah is the creator of all things."

"Where would I find him?"

"I cannot answer your question. I do not possess God's ability and his knowledge," Adam replies with sadness.

"Is your creation from God?"

"Yes, it is, in all that I am and all that I have."

"Adam, you will write about everything that occurs in your lifetime so the world's population will learn all that was, beginning with your life."

"I have one question," says Adam. "What is your name?"

"Primo, as yours is Adam."

Primo places his hand again on Adam's forehead.

Adam goes to sleep, and Primo departs from there. In a second, he is back to Cain's time at the point of transformation by the Others.

"This is the work of the Others, not of the creature that Adam claimed, not Lucifer, the evil mind taker." Primo's new task is to understand the power of Lucifer. For him it is becoming a wider task to find and uncover the means of it all.

He moves to AD 33. Again, he sees the man Jesus being crucified. Then he leaps back through all 33 years to unveil the life of Jesus. He knows all that he read in the writings of the Bible, but he find that it is just the history of a man and his birth.

Primo then travels to a time after Jesus' death and sees Jesus alive again.

"I do not understand this! Is Jesus a clone?" Primo becomes confused by what he finds out about Jesus.

For forty days, Primo will follow Jesus, to the day when he sees Jesus on a mountain top speaking to a crowd of people. There he observes Jesus flying up in slow motion over the clouds. Primo notes that Jesus looks at him with a serious smile, and then Jesus disappears into the cosmos in the blink of an eye. Primo travels as fast as he is capable through the cosmos and back to see if he can find where Jesus went. He wants to locate his home, his planet.

"This man God and his power are beyond my knowledge and understanding. Maybe my family and I are part of him the same as Adam."

Primo undertakes a new exploration to find the truth about the history of Jesus. He travels back to time past, where Jesus is departing from the crowd. He does the same investigation to the micro-millionth of a second to see if he can solve the mystery of Jesus and his departure. Every time is the same. There is no change in the event.

In seconds, Primo is once again in 2012.

"The writings of the book called the Bible are very advanced, too advanced for man to understand. It is complicated. It hides great knowledge. I will concentrate on the last book, which is called Revelation."

Primo is determined to find all that he can find to save the world and a dimensional time so man can leave and his planet will remain as it is.

Primo begins to read Revelation in every translation in every language. In seconds, Primo knows and learns the entire book, from chapter one to the end chapter.

Revelation is the last book of the New Testament. It tells of the end days and the judgment of mankind for all he has done to himself and to God's creation. Punishment will be harsh and terrifying.

RELIGIONS & CULTS

Primo find himself mind-boggled by the reading and understanding of this strange book. He knows that the end of man is not from God but from the Others.

"I will read all the religious writings from all cultures to see if they have at the same philosophical understand of the end of the human race and the world." We know that Primo is doing all this work to find out the truth so that his planet—the green planet—will stay safe.

Again, Primo goes back and forth according to the writing of man's knowledge about history and his inner beliefs. Through the ages, he finds different writings that speak of the events of the end times.

"This is strange and inaccurate. They all say the same thing, but the actual occurrences are different than their writings." He knows the truth and how history is made by man and his spiritual endeavors.

"From all the information I have uncovered and my findings, I will not change history for humanity."

For the last time, Primo travels back to Adam's time. Primo is convinced that he is the first created man, but he cannot find from where, so he stops the search. Primo is also convinced that Jesus is both human and God, but he cannot find Jesus and his home no matter how far he searches, so he stops the search in that regard as well.

The book of Revelation is unclear to him. So are all the other writings.

"In the present world, man will not have a solution for peace and tranquility. It will be only by a miracle that they believe."

Human theories are vast about God or gods and non-gods, just science through imaginations and hope. Primo does not find any truth, especially in cults and science. To him the writings are bogus and play into the greed and selfishness of mankind. He believes that man is confused, lost in his own mind, and has no hope for salvation. The end is inevitable.

"My commitment to the humans and the planet is carefully engaged. I must save what is right and valuable for the new Earth, before the end."

He selects and collects all that the Earth provides, from seeds to cells, DNA from all kinds of creatures from the sky, ocean, and land, all cellular material for life. When he is finished, he leaves Earth.

Using time travel, Primo departs the Earth and heads to the end of the Milky Way, the opposite side of the galaxy. He knows of a large planet in existence that will be suitable for human life and all the species from Earth.

"It will be a new Earth for the new human race, with good quality materials making this new Earth fertile."

He hovers over the new planet observing its surroundings. The new Earth has a moon and many other planets, with a star much larger than the Earth's sun.

"A perfect home."

From above the planet, Primo gets all the materials he collected and carefully selected. He places them together in a sphere the size of a basketball. In seconds, it flies down to the planet, causing an amazing nuclear explosion that causes a tsunami that goes around the whole planet. Primo observes the occurrence, and then departs to the current Earth.

Primo remains on the Earth for many years. He teaches man all good things for man to achieve, from fertilizers for the Earth to the study of nature and how to evolve in medicine and science. He is there from the time of Adam until 2022. He leaves signs of knowledge so man can uncover the truth about himself and signs about the Others.

Primo selects the best of human life from every continent, from the beginning of humans on Earth to before the end of human life. He calls upon his brothers, C=2 and C=3, to come and assist with the transformation and transportation of human life as it is collected by Primo, three percent of adults and seven percent of children. These are the only humans saved to live on the new Earth.

Now the three of them together place all the humans in a large sphere. The humans are in a state of suspended animation for the trip.

"I sense that these humans will maintain order with integrity on the new planet," Primo tells his brothers.

"You did right brother, in just a short time," the two brothers reply telepathically.

They depart Earth's atmosphere with all the humans at the end of time in 2040 and travel to the new Earth.

They arrive there in a short time. They descend to the new Earth. Primo and his brothers see the beautiful planet with all living things that Primo brought, things that make it possible for the humans to live.

"It is fantastic, Primo!" C=2 expresses with gladness.

"Yes, brother. They will live here in peace and harmony, and someday they will find their God."

"Do you want us to remain with you?" C=3 asks Primo.

"You shall return home. I will remain here for a few seconds to observe them, to make sure they will be fine without me checking their future."

They depart from one another. Primo is now hovering over the new Earth. He remains there for a time, and then returns to the old Earth in 2041. He sees the planet, scorched, supporting no life. Everything in the Earth's solar system is out of order.

"This motion will go on for seven million years."

Primo is sad for the Earth and everything that surrounds it, but he knows that the changes in the solar system will be his and his family's gain.

He leaves the Earth's solar system and goes to the new Earth. He stays there for a while, then goes back home his green planet. There he gets together with his family to report on the entire occurrence.

The clones that man made were for the purposes of power and glory. The limitation that God give man is infinity, all because of choices … yes, freedom of choices. The Strangers and the Others are the results of flaws in man, thinking and striving to be a god, making mistake after mistake. Even after Primo left signs of knowledge to develop the mind of man, the conflict between good and evil that comes from the Strangers and the Others will make man's ideas confused and disorganized. Humans will collaborate to develop plans, lying and deceiving one another.

In his conference with his family, Primo engages in a review of things regarding the human race, including their God and religions.

"The relics of religion and my understanding of them are cloudy. The humans are critiquing and denying the truth of everything. They live in lies." Primo express his knowledge to all the other clones.

"Now that the planet Earth is gone, soon to become the future green planet, our home, what will be our mission to the new Earth?" C=2 asks on behalf of all the clones.

"We will be the guardians of the cosmos for as long we exist, to help not only the new Earth but other planets with life. We also have a task to find the god named Lucifer. His power is to deceive and destroy all that is good by all means. Yes, we are the guardians of the cosmos."

The answer that Primo gave to all confirms that the clones will be traveling the universe forever. Time is essential to their life and work.

Are they safe? Or they will continue to be concerned about the Others? Will the Others come back to find the Strangers and battle them and other living creatures from other planets? Definitely! The Others are far, far away from the solar system, engaging in other acts of destruction.

But no matter where they are, eventually they will meet again, because evil and good will never disappear from the cosmos.

Epilogue

We come to the conclusion of *Clones: Strangers vs. The Others*. It contains many facts and idea from my studies and experiences in life about God and science. We know that science has tried to falsify God for generations. We know that creation was not done by mistake or by accident. DNA is designed, genes made of cells, chemical and nucleic material, to create all living things on planet Earth. People of science are like cats, curious to find the unknown and make something out of nothing. Ninety-nine percent is theory, and one percent is fact.

The human brain is a large organ with billions of cells, nerves, and electrical pulses. It continuously transmits information to the body at extraordinary speeds. No other creature even comes close to the human brain.

The human brain is a gift from God. If the brain dies, the body dies.

Some people are misled by science, which encourages weak thinkers to dismiss God from all earthly matters ... from dust, seeds, plants, water, minerals, and all life that God created.

Man is stupid. His ignorance is vast. He is a fake and disingenuous with himself and others. His strange ideas convert people with problems and lack of knowledge. Most people will deny God so they can test a bogus life of knowledge and information. They will believe that man comes from fish to ape/monkeys, that man existed for over 100,000 years on this, our planet Earth. Some people make themselves believe that God is just an idea and Jesus is a character. They think the Bible is mythical, like the gods in fables … just stories.

Science makes us believe that out in space are ETs and aliens with intelligence superior to that of humans. They can travel at the speed of light. That is laughable, a determination of knowledge from ideas and theory, by people who know that they are not convinced about life in outer space. No, they are not sure.

The saddest part of all is the lies and bogus information accepted by the average person who reads science books, watches television, and listens to radio reports. If a blind man leads a blind man, they both will fall in the pit.

I remind you that I like science. I enjoy new discoveries and I feel that God made the universe and the protection of Earth for us. In the first chapter of the book of Genesis, it tells us that after He completed the Earth and all living things, including man, then he created the universe.

If you do not believe me, read. Get the truth and facts from the real source—God. He is the one who made you and everything else, including science.

May God help you in your search for the truth as you endeavor to gain knowledge and wisdom in your new discoveries.

Why do I invoke the name of God in this science fiction novel? Because it goes with it. We cannot separate God from all we do, especially science, mathematics, and medicine. Science will never convince me that we are made from the big bang or universal dust from a meteorite. We have all the information that we need to learn about God and creation. What we need to do is written and mapped very clearly. That is not fiction; it is the truth. The universe, the cosmos, is created with fine-tuned precision, designed in micro-calculations with no flaws, engineered to the core. Only a perfect being would create such an incredible thing. Man will never understand the almighty power of God.

Check your IQ. In the number grid on the next page, find a square and a pyramid with four triangles through the binary mathematic table in calculus of seven.

Binary
Mathematics
Hexadecimals
Equations Bits

In mathematics and science, the binary numeral system (base 2) represents numeric values using two symbols: 0 and 1. More specifically, the usual base 2 system is a positional notation with a radix of 2. Numbers represented in this system are commonly called binary numbers. Because of its straightforward implementation in digital electronic circuitry using logic gates, the binary system is used internally by almost all modern computers and computer-based devices.

```
1 0 0 1 0 1 1 1 1 0 1 0 1 0 0 0 0 1 0 1 0 0
1 0 1 0 1 0 0 1 0 1 0 0 1 0 1 0 0 1 0 1 0 0
1 0 0 1 0 1 1 1 1 0 1 0 1 0 0 0 0 1 0 1 0 0
1 1 1 0 0 0 0 1 0 1 0 1 0 0 1 0 1 0 0 1 1 0
1 0 0 0 1 1 0 1 1 0 0 1 0 0 0 1 0 1 0 0 1 1
1 0 1 0 0 1 1 1 0 1 1 0 0 0 1 0 0 1 0 1 0 0
1 1 1 0 1 0 0 1 1 0 0 1 0 1 0 1 1 0 0 1 1 0
0 0 0 0 1 1 1 1 0 0 0 1 0 1 0 0 1 0 0 1 1 0
0 1 0 1 0 0 0 0 0 1 0 1 1 1 1 1 0 1 0 1 0 0
1 1 1 0 0 0 1 1 1 0 0 0 1 0 1 1 1 0 0 1 1 0
1 0 0 0 1 1 1 0 0 1 1 0 1 0 0 0 1 0 0 0 1 1
1 0 1 0 1 0 0 1 1 1 0 0 0 1 1 0 0 0 1 0 0
1 1 1 0 1 1 1 0 0 1 0 0 1 1 0 1 0 0 0 1 1 0
1 0 0 1 0 0 0 0 0 1 1 0 1 1 1 0 1 1 0 1 0 1
1 1 0 0 0 1 0 0 1 0 1 1 1 0 0 1 0 1 0 1 0 0
1 1 1 0 0 0 1 0 1 0 1 0 1 1 1 0 0 1 0 1 1 0
0 0 0 0 1 0 0 1 1 0 0 1 0 1 0 0 1 1 0 1 1 0
1 0 1 0 0 1 1 0 0 1 1 1 0 0 1 0 1 0 0 1 0 0
1 1 1 0 0 0 0 1 0 1 0 1 0 0 1 0 1 0 0 1 1 0
1 0 0 0 0 0 0 1 0 0 1 0 1 0 0 1 0 0 0 0 1 1
1 0 1 0 1 0 0 1 0 1 0 0 1 0 1 0 0 1 0 1 0 0
1 1 1 0 0 0 0 1 0 1 0 1 0 0 1 0 1 0 0 1 1 0
```

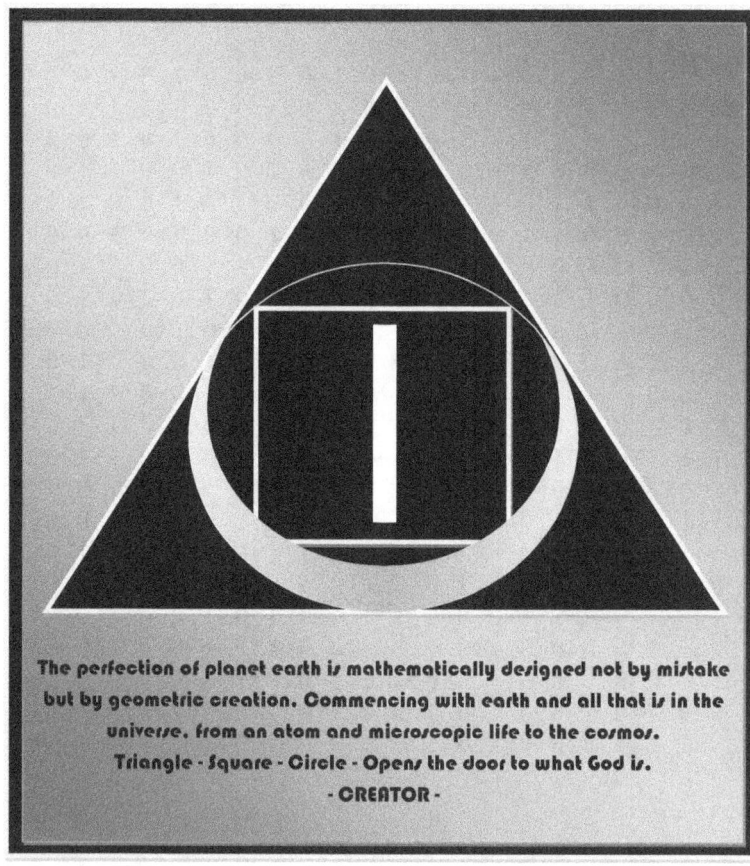

The perfection of planet earth is mathematically designed not by mistake but by geometric creation. Commencing with earth and all that is in the universe, from an atom and microscopic life to the cosmos. Triangle - Square - Circle - Opens the door to what God is.
- CREATOR -

Acknowledgments

The All-Mighty God: God is truth, and the truth cannot be changed. My greatest gratitude goes to God and my Lord Jesus Christ for everything I do well in life. If it were not for Him, I would not be here.

I thank my parents and family for being kind and loving people.

This book is dedicated to the living memory of my mother, Gaetana Masi, and my uncle, Francesco Masi, for their great doctrine and philosophical input in my life.

Last—but not least—I thank you, the reader, for buying this book. I hope you enjoy it. Please look for other fantastic and intriguing action and science fiction novels, comics, and more from M.R. Comics & Art.

For more information contact:
    M.R. Comics & Art
    www.mrcomicsandart.com
    info@mrcomicsandart.com

Art by Giovanni Rocca available through www.mrcomicsandart.com

Previous books in the Clones Trilogy

Available from MR Comics & Art and Amazon.com

coming soon

Purchase your short run, first print *Accor: Vendetta* package. Only 100 are available, each containing a certificate of authenticity, along with numbered and signed copies of *Accor* nos. 1, 2, 3, and 4 by Giovanni Rocca and Luke McDonnell, plus *Accor* no. 1 variant signed painted cover and a pin up poster by Luke McDonnell. All for $49.99 (shipping and handling included). Email to reserve your package today at info@mrcomicsandart.com.

Screenplay

By

Giovanni Rocca

Children Of the Sun

M.R. Comics & Art   Giovanni Rocca
                           info@mrcomicsandart.com

ED WILD ANIMALS - EARLY AFTER

the air. Through the
d appears. On the top of the
ife aims at the sun, praying
by sacrificing a human

e crowd below around the
ifully decorated with
attoos; heads and garments
er stones and feathers of

stone altar; four men are

ec)
lorious day
...so you may
ounty...For
y existence!

the chest, cutting his heart
another priest. The other
es it on a firing plate, to
ring to please the gods.

god!

ng down from the top of the
utcome.

The King sits on his throne next to him his Queen; the King
stands and looks at the crowd with gladness. With a burst of
energy and power he glorifies the occurrence.

                    KING
               (shouting)
          My people! Children of the
          sun...Today once more the gods are
          content, and give us life again!

                                        (CONTINUED)

Children of the Sun

124 page screenplay
available to producers,
directors, actors, and agents.
M.R. is looking to hire agents.

To inquire, email M.R. Comics & Art at info@mrcomicsandart.com

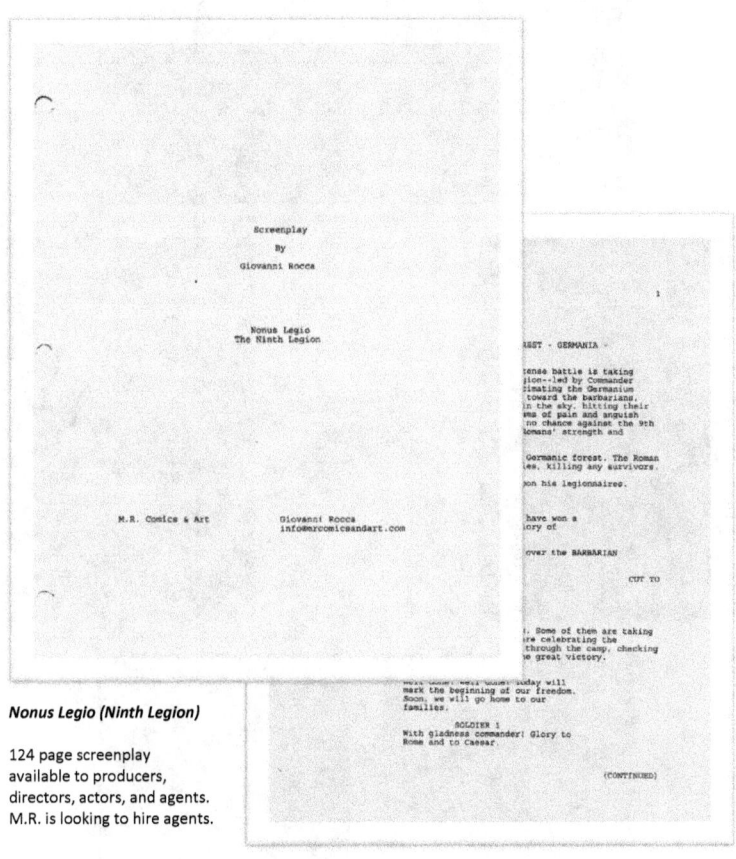

Comics and Graphic Novels